Eadweard Muybridge

Descriptive Zoopraxography

The science of animal locomotion made popular

Eadweard Muybridge

Descriptive Zoopraxography
The science of animal locomotion made popular

ISBN/EAN: 9783337235550

Printed in Europe, USA, Canada, Australia, Japan

Cover: Foto ©berggeist007 / pixelio.de

More available books at **www.hansebooks.com**

DESCRIPTIVE
ZOOPRAXOGRAPHY

OR THE SCIENCE OF ANIMAL LOCOMOTION MADE POPULAR

BY

EADWEARD MUYBRIDGE

WITH SELECTED OUTLINE TRACINGS REDUCED FROM SOME OF
THE ILLUSTRATIONS OF

"ANIMAL LOCOMOTION"

AN ELECTRO-PHOTOGRAPHIC INVESTIGATION OF CONSECUTIVE
PHASES OF ANIMAL MOVEMENTS, COMMENCED 1872,
COMPLETED 1885, AND PUBLISHED 1887,
UNDER THE AUSPICES
OF THE

UNIVERSITY OF PENNSYLVANIA

PUBLISHED AS A MEMENTO OF A SERIES OF LECTURES GIVEN BY THE AUTHOR
UNDER THE AUSPICES OF THE UNITED STATES
GOVERNMENT

BUREAU OF EDUCATION

AT THE

WORLD'S COLUMBIAN EXPOSITION, IN ZOOPRAXOGRAPHICAL HALL

1893

University of Pennsylvania
1893

SOME OF THE SUBSCRIBERS

TO

"ANIMAL LOCOMOTION."

———————

THE ORIGINAL AUTOGRAPHS ARE ON THE SUBSCRIPTION BOOK
IN THE POSSESSION OF THE AUTHOR.

———————

T. H. Huxley E. Ray Lankester

W. H. Flower John Lubbock M. Gaudry

Richard. Owen A. Günther

G. J. Romanes G. Max Müller

Rudolf Virchow L. Pasteur

A. Agassiz Joseph Leidy

Francis Galton O. C. Marsh W. Wundt

Paolo Mantegazza St George Mivart

Alfred Newton F. A. Jensen

H. C. Sorby W. Dönitz K. Zittel

G. Brown Goode, G. Barrier Max Weber.

F. Blochman K. Brandt.

J. S. Billings Theo. Gill

A. Milne Edwards A. B. Meyer.

C. E. Brown-Séquard Rückert

Henry W. Acland F. Darwin

J. Bell Pettigrew H. le Wood

Wm. H. Brewer J. Mowes

C. Ludwig E. Pflüger E. du Bois-Raymond

M. Foster.

J. Rosenthal. R. Heidenhain Fr. Goltz

Sigm. Exner A Mosso

H. P. Bowditch M. Schiff

J. Burdon Sanderson A. Fick

Moleschott E. Hering

Pietro Albertoni H. Aubert.

Weir Mitchell

William Pepper N. Kintschgau

C. v. Voit

J. v. Kries. Hermann Munk.

Wilh. Braune A. Kölliker L. Landois.

John Cleland D. Hayem Agnew.

G. M. Humphry.

John Marshall Carl Rabl.

C. Hasse. J. P. Gruetzner. Müllenhoff

W. Kühne J. Bernstein Meissner

G. Stieda Heusen W. Flemming

Ph. Stöhr C. Slosse J. Miescher

Luigi Luciani Julius Gaule

Edouard Van Beneden E. Masoin

H. v. Helmholtz   H. W. Vogel

William Thomson   S. P. Langley

H. F. Weber.   E. Mach.

Frederick Bramwell

Thomas A Edison   W. Spottiswoode

W. de W. Abney   R. W. Bunsen.

William Huggins   G. Mattuing?g

M. Bellati

W. C. Ch. Soret   Maserna?

Foerster Pietro

L. Ditscheiner, Hagenbach-Bischoff

Gaston Tissandier

James Glaisher

E. Rousseau

Francis Blake

Wm. A. Wahl

W. W. Crump

Franklin Leonard Pope

Antonio Roiti Georg Quincke

Wm.

C. A. Juling Sellers

Augusto Righi   Irving M. Scott

G. F. Watts

Hubert Herkomer
Edwin Long
Thomas Faed
W. Holman Hunt
C. S. Markes
Luke Fildes
W. W. Ouless
J. McNeil Whistler
E. Onslow Ford
George du Maurier
Lucas
Alfred J. Hicks J. Edgar Boehm
Edmund J. Poynter
John Everett Millais
John Severn
W. Q. Orchardson
G. A. Storey

LIST OF SUBSCRIBERS

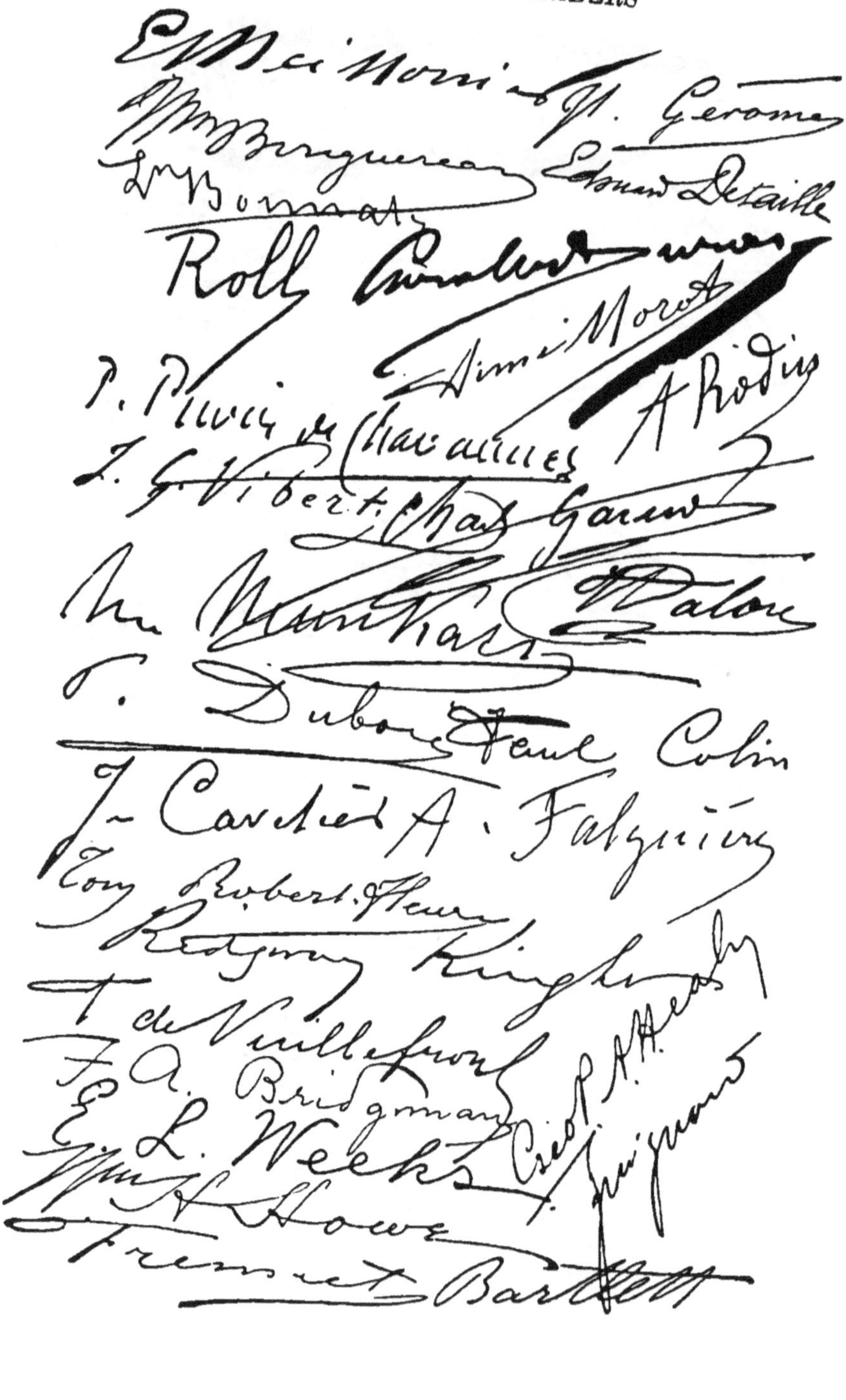

Menzel

Werner

F. Lenbach   Ludwig "Knaus".

Carl Becker.   Chr. Ruß

R. W. Wagner

Beg...   Gab Max

   C. V. Gysis

   R. Siemering

Schaper

Paul Meyerheim   Müller

Eugen   von   Anton Heß

H. Ernst   Carl Marr

   F. von Uhde

...   Thiersch

...   Johannes Schilling

Wilh. Sils   P.   Janssen

Wil. Riefstahl   Eduard Grützner

H. Zügel   C. Willr.   Diez

   Anton Braith

   Hübner   F. Miller

F. Av. Kaulbach

(page of handwritten signatures)

Ludwig Passini

Tito Conti   W. W. Story

Francesco Vinea   Hebert

Elihu Vedder   Sommer

Morelli   José Gallegos

Michele Gordigiani

Edoardo Gelli   Hopf

Nunzio Barbino

Augusto Bompiani

Giovanni Costa   Carl Lotz

Lingworth Powers   Székely Bertalan

Diego Sarti   Benvenuti

Charles Caryl Coleman

Salvatore Albano

Onorato Carlandi

Enrico Coleman

H. H. Richardson

J. L. award

Sarah W. Whitman

Louis C. Tiffany

J. G. Brown

Launt. Thompson jr

B. Uhle

Aug. St. Gaudens

William T. Richards

Thos. Moran

Eastman Johnson

Ernest. W. Longfellow

Anne Whitney

Jos Keppler

Geo B Post

Hamilton

R. Swain Gifford

Jno LaFarge

Thomas Allen

J Morris

E woods Perry Jr

Edw H Kendall

F. D. Millet

D. H. Burnham.

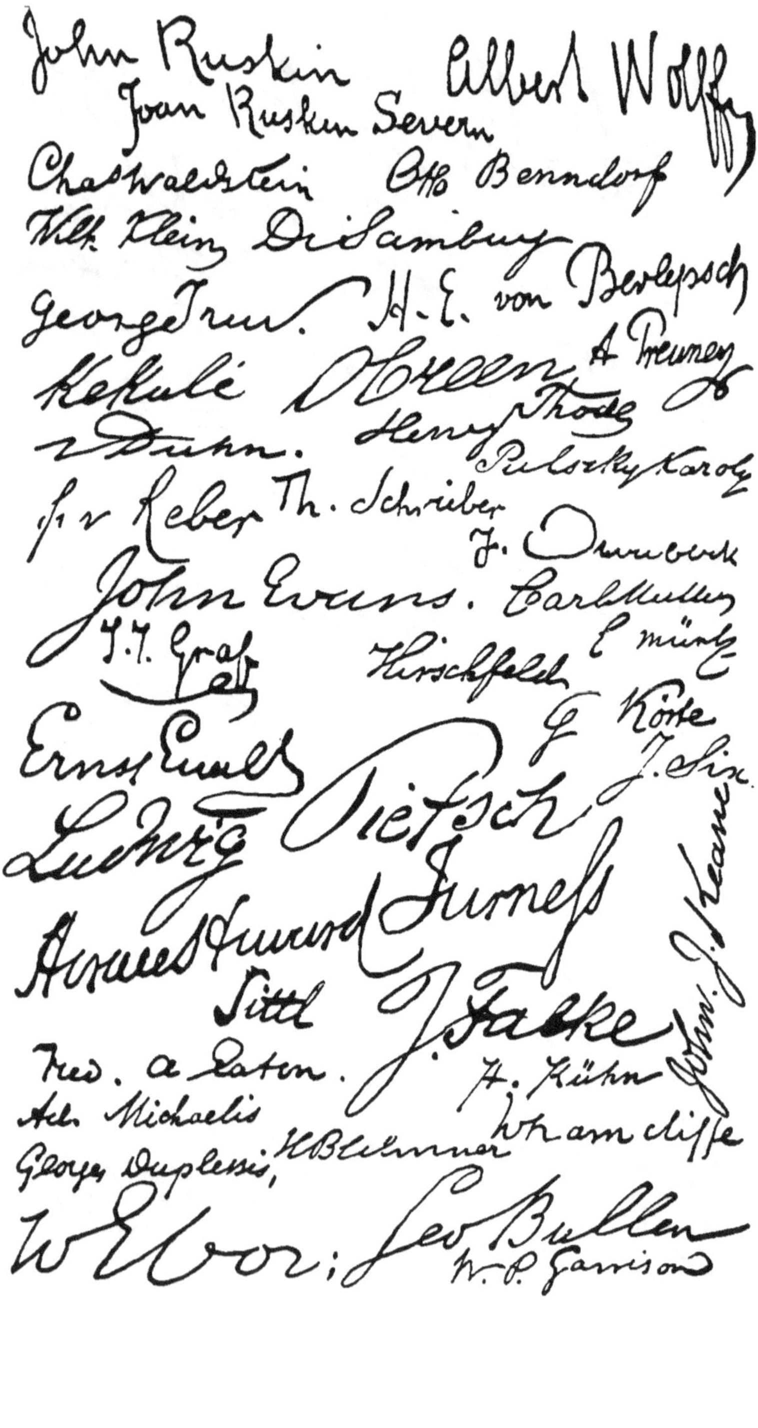

[Page of handwritten signatures of subscribers]

John Ruskin   Albert Wolff

Joan Ruskin Severn

Chas Waldstein   Otto Benndorf

Wilh. Klein   Di Samburg

George True   H. E. von Berlepsch

Kekulé   Green   A. Preuner

v. Duhn.   Henry Thode

Pulszky Karoly

f. v. Reber   Th. Schreiber

F. Dümert

John Evans.   Carl Müller

J. J. Graf   Hirschfeld   E. münz

G. Körte

Ernst Curtius   Pietsch   J. Six

Ludwig

Horace Howard Furness

John J. Keane

Sittl   J. Falke

Fred. a. Eaton.   F. Kühn

Ad. Michaelis   H. Blümner   Wh. am cliffe

Georg Duplessis,

W. E. Cox;   Geo. Bullen

W. P. Garrison

Bertha M. H Palmer

C. Vanderbilt

Pierpont Morgan

a Drexel D O Mills

Chas. C. Harrison

Martin A Ryerson.

C L Hutchinson

Geo. W. Childs

Wharton Barker

Marshall Field

Saml Dickson

K L

Craige Lippincott

Fred. L. Ames.

Jos. W Drexel

Edward H Coates

Wm Weightman

Thomas Hockley

Francis

H. H. Houston

Sully Darley

W J Walters

John H Converse

Geo. W. Vanderbilt

A. Biddle

Rutherfurd Stuyvesant

L H: Hauffmann.

T. Jefferson Cooledge.

PREFACE.

In the summer of 1892 while the Author was in California, preparing for a Lecturing tour through Australia and India, he received an invitation from the Fine Arts Commission of the World's Columbian Exposition to give a series of Lectures on ZOOPRAXOGRAPHY in association with the Exposition now being held in Chicago.

As these Lectures under the more familiar title of "The Science of Animal Locomotion in Its Relation to Design in Art" had already been given at nearly all the principal Institutions of Art, Science and Education in Europe and in the United States, (see appendix A) the Author was induced to believe that they might be repeated in a popular manner at the Exposition, with

1

some appreciation of the importance of the facts which his investigation has revealed, not merely by the student of Nature or of Art, but by that large and important class of students, known as the general public. Under this impression he delayed his far Occidental expedition and returned to Chicago to find a commodious theater erected for this special purpose on the grounds of the Exposition, to which the name of Zoöpraxographical Hall had been given; the Science of Zoöpraxography having had its origin in the Author's first experiments in 1872. It is not intended in this monograph to give more than a synopsis of the usual course of Lectures on the subject, nor to reproduce any of the pictured or sculptured representations which are necessary for its proper elucidation, but merely to describe the common methods of limb action adopted by quadrupeds—especially by the horse—in their various acts of progressive motion, and to illustrate the most important phases of these movements by tracings from the original photogravures of the Author's work.

In the presentation of a Lecture on Zoöpraxography the course usually adopted is to project, much larger than the size of life upon a screen, a series of the most important phases of some act of animal motion—the stride of a horse, while galloping for example—which are analytically described. These successive phases are then combined in the Zoöpraxiscope, which is set in motion, and a reproduction of the original movements of life is distinctly visible to the audience.

With this apparatus, horse-races are reproduced

with such fidelity that the individual characteristics of
the motion of every animal can readily be seen; flocks
of birds fly across the screen with every movement of
their wings clearly perceptible; two gladiators con-
tend for victory with an energy which would cause the
arena to resound with wild applause, athletes turn
somersaults, and other actions by men, women and
children, horses, dogs, cats and wild animals, such as
running, dancing, jumping, trotting and kicking, are
illustrated in the same manner. By this method of
analysis and synthesis the eye is taught how to ob-
serve and to distinguish the differences between a true
and a false impression of animal movements. The
Zoöpraxiscopical exhibition is followed by illuminated
copies of paintings and sculptures, demonstrating how
the movement has been interpreted by the Artists of
all ages; from the primitive engravers of the cave
dwelling period, to the most eminent painters and
sculptors of the present day.

INTRODUCTION.

In the year 1872, while the Author was engaged in his official duties as Photographer of the United States Government for the Pacific coast, there arose in the city of San Francisco one of those controversies upon Animal Locomotion, which has engaged the attention of mankind from the dawn of symbolical design, to the present era of reformation in the artistic expression of animal movements.

The subject of this particular dispute was the possibility of a horse having all of his feet free of contact with the ground at the same instant, while trotting, even at a high rate of speed, and the disputants were Mr. Frederick MacCrellish and the Hon. Leland Stanford.

The attention of the Author was directed to this controversy and he immediately sought the means for its settlement.

At this time the rapid dry plate had not yet been evolved from the laboratory of the chemist, and the problem before him was to develop a sufficiently intense and contrasted image upon a wet collodion plate, after an exposure of so brief a duration that a horse's foot moving with a velocity of more than a hundred lineal feet in a second of time, should be photographed practically "sharp."

A few days' experimenting and about a dozen negatives, with a celebrated fast trotter—"Occident"—as a model, while trotting at the rate of a mile in two minutes and sixteen seconds, laterally in front of the camera, decided the argument for once and for all time in favor of those disputants who held the opinion that a horse while trotting was for a portion of his stride entirely free from contact with the ground. With a knowledge of the fact that some horses while trotting will make a stride of twenty feet or more in length, it is difficult to understand why there should ever have been any difference of opinion on the subject.

These first experiments of Zoöpraxography were made at Sacramento, California, in May, 1872. A few impressions were printed from the selected negative for private distribution, and were commented upon by the "Alta California," a newspaper published in San Francisco.

Thus far the photographs had been made with a single camera, requiring a separate trotting for each exposure. The horse being of a dark color and the background white, the pictures were little better than silhouettes, and it was difficult to distinguish, except by inference, the right feet from the left.

Several phases of as many different movements had been photographed, which the Author endeavored with little success to arrange in consecutive order for the construction of a complete stride.

It then occurred to him that if a number of cameras were placed in a line, and exposures effected successively in each, with regulated intervals of time or of distance, an analysis of one single step or stride could be obtained which would be of value both to the Scientist and the Artist.

The practical application of this system of photo-

graphing required considerable time for its development, and much experimenting with chemicals and
apparatus.

It being desirable that the horses used as models
should be representatives of their various breeds, and
the Author not being the owner of any that could be
fairly classed as such, obtained the coöperation of
Mr. Stanford, who owned a fine stud of horses at his
farm at Palo Alto, and there continued his labors.

The apparatus used at this stage of the investigation was essentially the same as that subsequently
constructed for the University of Pennsylvania, the
arrangement of which will be described further on.

Some of the results of these early experiments which
illustrated successive phases of the action of horses
while walking, trotting, galloping, &c., were published
in 1878, with the title of "The Horse in Motion."
Copies of these photographs were deposited the same
year in the Library of Congress at Washington, and
some of them found their way to Berlin, London,
Paris, Vienna, &c., where they were criticized by the
journals of the day.

In 1882 the Author visited Europe and at a reception given him by Monsieur Meissonier was invited by
that great painter to exhibit the results of his labors to
his brother Artists who had assembled in his studios for
that purpose. M. Meissonier was the first among
Artists to acknowledge the value to Art design of the
Author's researches; and upon this occasion, alluding
to a full knowledge of the details of a subject being
necessary for its truthful and satisfactory translation
by the Artist, he declared how much his own im-

pression of a horse's motion had been changed after a careful study of its consecutive phases.

It is scarcely necessary to point out, in confirmation of M. Meissonier's assertions, the modifications in the expression of animal movements now progressing in the works of the Painter and the Sculptor, or to the fact of their being the result of studious attention to the science of Zoöpraxography.

In the same year, during a lecture on "The Science of Animal Locomotion in Its Relation to Design in Art," given at the Royal Institution (see *Proceedings* of the Royal Institution of Great Britain, March 13, 1882), the author exhibited the results of his experiments at Palo Alto, when he, with the Zoöpraxiscope and an oxy-hydrogen lantern, projected on the wall a synthesis of many of the actions he had photographed.

It may not be considered irrelevant if he repeats what he on that occasion said in his analysis of the quadrupedal walk:—

"So far as the camera has revealed, these successive foot fallings are invariable, and *are probably common to all quadrupeds.* . . . .

"It is also probable that these photographic investigations—which were executed with wet collodion plates, with exposures not exceeding in some instances the one five-thousandth part of a second—will dispel many popular illusions as to the gaits of a horse, and future and more exhaustive experiments, with the advantages of recent chemical discoveries, will completely unveil all the visible muscular action of men and animals even during their most rapid movements. . . .

"The employment of automatic apparatus for the

purpose of obtaining a regulated succession of photographic exposures is too recent for it to be generally used for scientific experiment or for its advantages to be properly appreciated. At some future time the philosopher will find it indispensable for many of his investigations."

The great interest manifested in the results of his preliminary labors convinced the Author that a comprehensive and systematic investigation with improved mechanical appliances, and newly-discovered chemical manipulations, would demonstrate many novel facts, not only interesting to the casual observer, but of indisputable value to the Artist and to the Scientist. This investigation and the subsequent publication in the elaborate manner determined upon, assumed such imposing proportions, and necessarily demanded so large an expenditure, that all publishers, not unnaturally, shrank from entering the unexplored field.

In this emergency, through the influence of its Provost, Dr. William Pepper, the University of Pennsylvania with an enlightened exercise of its functions as a contributor to human knowledge, instructed the Author to make, under its auspices, a comprehensive investigation of "Animal Locomotion" in the broadest significance of the words, (see appendix B) and some of the Trustees and friends of the University constituted themselves a committee for the purpose of promoting the execution of the work. These gentlemen were Dr. William Pepper, Chas. C. Harrison, J. B. Lippincott, Edw. H. Coates, Samuel Dickson and Thomas Hockley.

The Author acknowledges his obligations to these gentlemen for the interest they took in his labors; for without their generous assistance the work would probably never have been completed; the total amount expended—nearly forty thousand dollars—being entirely beyond his own resources. To Drs. F. X. Dercum, Geo. F. Barker and Horace Jayne, of the University, the Author is also indebted for much valuable assistance.

Diagram of the Studio at The University of Pennsylvania, and Arrangement of the Apparatus for Investigating Animal Locomotion.

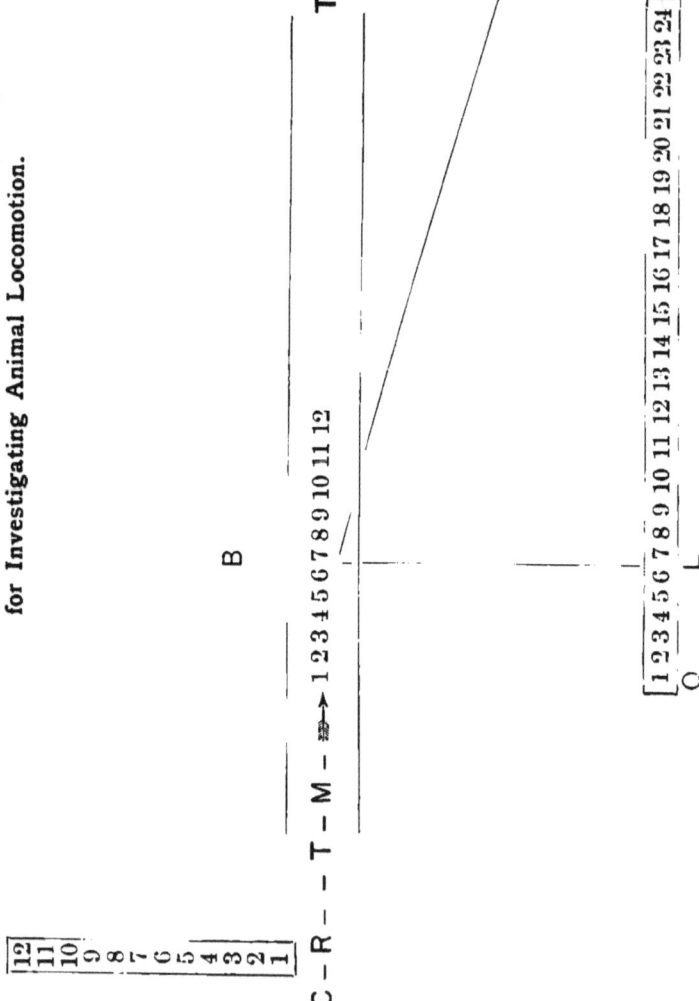

STUDIO, APPARATUS, AND METHOD OF WORKING.

For a proper appreciation of the care taken in the Investigation of Animal Locomotion at the University of Pennsylvania to ensure accurate record of the consecutive phases of the various movements, attention to the system adopted is necessary.

In the diagram, B is the *Lateral* background; consisting of a shed 37 metres or about 120 feet, long, the front of which is open, and divided by vertical and horizontal threads into spaces 5 centimetres, or about 2 inches, square, and by broader threads into larger spaces 50 centimetres, or about 19¾ inches, square.

At C and C, 37 metres, or about 120 feet, apart are "*fixed*" backgrounds, with vertical threads 5 centimetres, or about two inches, from their centres, with broader threads 30 centimetres, or about 12 inches, from their centres.

For some investigations, readily distinguishable in the plates, "*portable*" backgrounds are used, consisting of frames 3 metres wide by 4 metres high,—about 10 feet by 13 feet 4 inches,—over some of which black cloth and over others white cloth is stretched, all being divided by vertical and horizontal lines into square spaces of the same description as those of the lateral background.

These portable backgrounds are used when photo-

11

graphing birds and horses, and also wild animals when possible to do so.

L. A lateral battery of 24 automatic electro-photographic cameras, arranged parallel with the line of progressive motion, and usually placed therefrom about 15 metres or 49 feet.

Slow movements are usually photographed with lenses of 3 inches diameter and 15 inches equivalent focus; the centres of the lenses being 15 centimetres, or about 6 inches, apart.

Rapid movements are usually photographed with a *portable* battery of cameras and smaller lenses.

The centre, between lenses 6 and 7, is opposite the centre of the track T.

For illustrations comprising both "Laterals" and "Foreshortenings," cameras 1 to 12 only are used.

When "Laterals" alone are required, cameras 13 to 24 are connected with the system and used in their regular sequence.

R. A portable battery of 12 automatic electro-photographic cameras, the lenses of which are $1\frac{1}{4}$ inches diameter and 5 inches equivalent focus; the lenses are arranged $7\frac{1}{2}$ centimetres, or about 3 inches, from their centres. When the battery is used vertically, lens 6 is usually on the same horizontal plane as the lenses of the lateral battery.

In the diagram this battery is arranged *vertically* for a series of "Rear Foreshortenings," the points of view being at an angle of 90 degrees from the lateral battery.

F. A battery of 12 automatic electro-photographic cameras, similar to that placed at R, arranged horizontally for "Front Foreshortenings," the points of view averaging an angle of 60 degrees from the lateral battery.

O. The position of the operator; the electric batteries; the chronograph for recording the intervals of time between each successive exposure; the motor for completing the successive electric circuits, and other apparatus connected with the investigation.

T T. The track parallel with the lateral battery and covered with corrugated rubber flooring.

M. The model, approaching the point number "1" on the track where the series of photographic illustrations will commence.

An estimate having been made of the interval of time which will be required, between each photographic exposure, to illustrate the complete movement, or that portion of the complete movement desired, the apparatus is adjusted to complete a succession of electric circuits at each required interval of time, and the motor is set in operation. When the series is to illustrate *progressive* motion; upon the arrival of the model at the point marked "1" on the track, the operator, by pressing a button, completes an electric circuit, which immediately throws into gearing a portion of the apparatus hitherto at rest. By means of suitably arranged connections, an electric current is transmitted to each of the 3 cameras marked "1" in the various batteries, and an exposure is simultaneously made on each of the photographic plates, respectively, contained therein. At the end of the predetermined interval of time, a similar current is transmitted to each of the cameras marked "2," and another exposure made on each of the 3 next plates, and so forth until each series of exposures in each of the three batteries is completed. Assuming the operator to have exercised good judgment in regulating the speed of the apparatus, and in making the first electric contact at the proper time, and that the figures 1 to 12 represent the distance traversed by the model in executing the movement desired, the first three photographic exposures — that is, one exposure in each battery — will have been synchronously made when the model was passing the position marked "1" on the track T; the second three exposures will have been made when the model was passing the position marked "2," and so on until twelve successive exposures were simultaneously made in each of the three batteries. This perfect uniformity of time, speed, and distance, however, was not always obtained.

When this monograph was commenced it was not intended by the author to give any more than a general idea of the method adopted for obtaining the results of his investigation; it has, however, been considered that a few illustrations and brief description of the apparatus devised and used by him may not be without interest to other students.

For the use of these illustrations he is indebted to the courtesy of Rev. Jesse Y. Burk, the Secretary of the University, and to J. B. Lippincott Company, the publishers of "The Muybridge Work at the University of Pennsylvania," a book which contains, among other essays upon the subject, "Materials for a Memoir on Animal Locomotion, by Harrison Allen, M. D.," and "A Study of Some Normal and

Abnormal Movements, by Francis X. Dercum, M.D.,
Ph.D."

Figure 1 is a view of the building containing the
lateral battery of twenty-four photographic cameras,
all of which were used when as many consecutive
phases of an act of motion were required.

Immediately in front of each of these cameras, and
detached therefrom, was placed an electro-photographic
exposor, a side section of which is represented by
Figure 2, in which A is a continuous band of thin

Fig. 1.

rubber cloth impervious to light; the edges of which
are bound with strong tape, and arranged to run in a
groove, and over two rollers RR which are attached to
a frame.

In this endless band are two apertures OO of suit-
able size, and so arranged that their full openings as
they pass each other shall simultaneously take place in
front of the center of the lens L.

The upper and lower edges of these apertures are
kept taut by light steel rods attached to the tape binding.

To the lower rod of the front aperture is fastened a ring C and a cleat, to which some elastic rubber bands B are attached; these bands are easily removable and their number increased at discretion; in some instances of rapid exposures a tension of twenty-five pounds or more was required. On a shelf of the frame is a magnet M, over the top of which is arranged a steel lever G pivoted near the end D which terminates with a slightly indented projection.

The armature of the magnet is pivoted at H; its upper arm terminates with a shoulder I. S is a spring to prevent the accidental shifting of the shoulder from its contact with the lever when the exposor is ready for its function. N is a set screw to adjust the distance of the armature from the magnet. To prepare for a series of photographic exposures — the plates having been already placed in the cameras — the end of the lever G is placed under

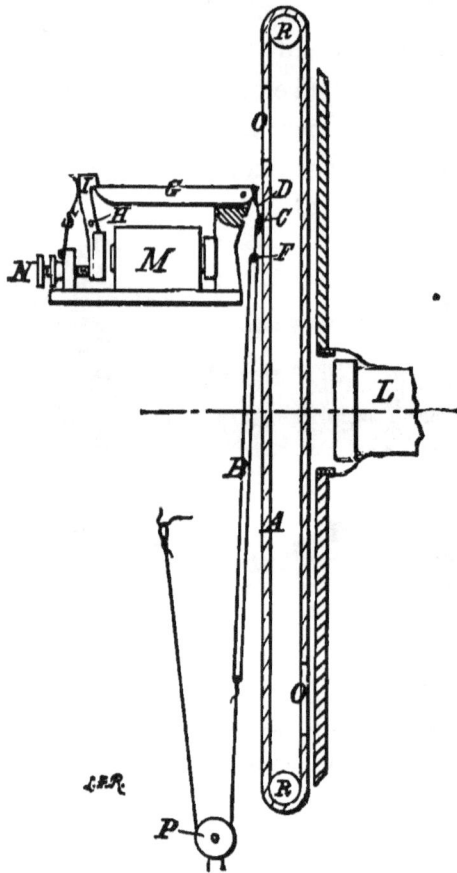

Fig. 2.

the shoulder I; the endless curtain is revolved until the front aperture O is raised to its proper position, when the ring C is hooked upon the projecting point D. A cord attached to the rubber bands B is drawn around the pulley P, and a ring at its end is slipped over a pin, which keeps the spring at a proper state of tension. Upon the completion of an electric circuit the armature is drawn towards the magnet; the end of the lever is released from its contact with the shoulder; the ring C is released from the projecting point D; the front of the endless curtain is drawn rapidly downward; the apertures meet in the center of the lens, form a gradually expanding and then contracting diaphragm, and the exposure is made. A front view

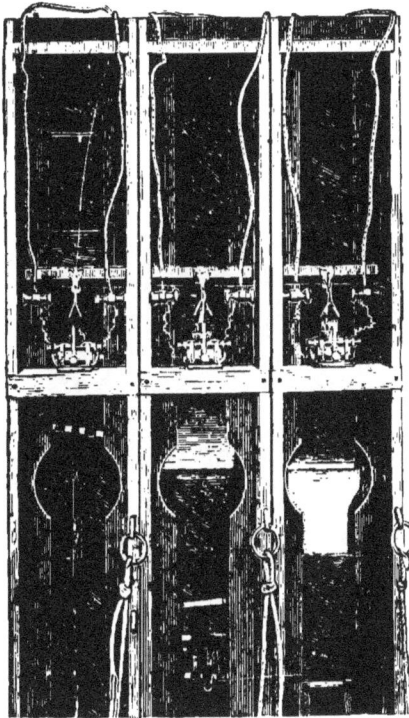

Fig. 3.

of three electro-photographic exposors is seen in Figure 3. The first of these represents the exposor set and ready for an exposure; the second shows the meeting of the apertures at the commencement of an exposure; the third, their position near the completion of the exposure, they having in the meanwhile uncovered the lens to their full capacity.

Figure 4 illustrates a portable battery of twelve electro-photographic exposors; it consists of a rectangular box divided into compartments, open at the front and rear.

In twelve of these compartments are arranged rollers, curtains, magnets, etc., as previously described, and a compartment through which a focusing lens is used. The two end compartments provide for the adjustment of the camera, which is supported in the box to the rear of the exposing arrangements. A

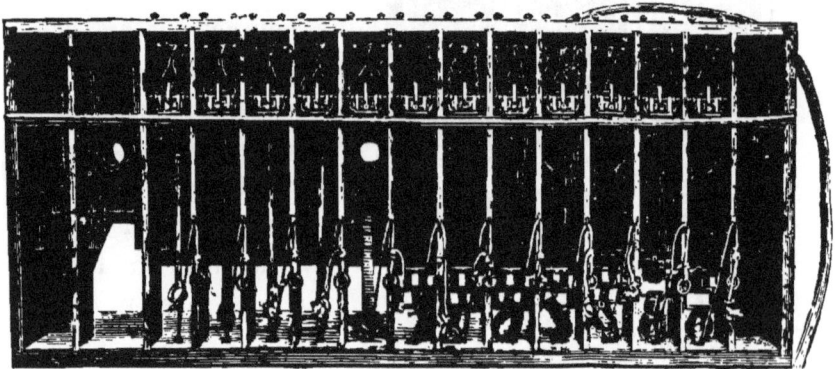

Fig. 4.

cable of insulated wires for connecting the twelve magnets with the exposing motor, contains a wire for the return current. As seen in the engraving, seven of the magnets by the passage of their respective currents have completed their releasing operations. In the eighth compartment the two apertures in the exposing band are in the act of effecting an exposure. The remaining four magnets are awaiting their turn for action.

Figure 5 is a photographic camera divided into

thirteen compartments, each having a lens of the same construction, and the same focal length; these are arranged to correspond with the compartments in the electro-exposors.

One of the lenses is provided with a focusing screen, and with it the other twelve lenses are adjusted to a proper focus without removing the plate holder behind them from its position in the camera.

The plate holder is constructed to hold three dry plates, each three inches by twelve inches; the front is divided into twelve compartments, each three inches square.

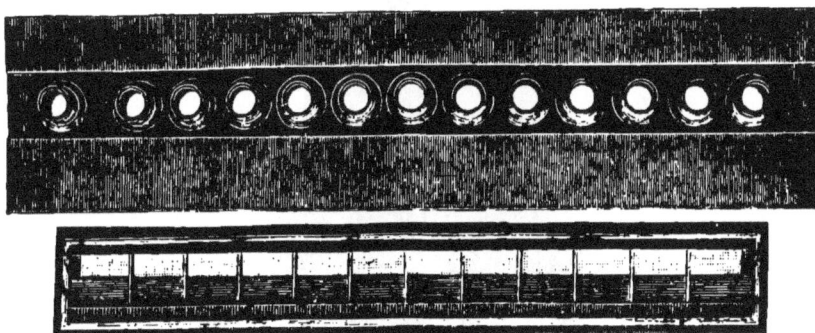

Fig. 5.

Light is excluded from the front by a roller blind, strengthened by thin narrow slats of hard wood; the blind works in grooves, is drawn over a concealed roller, and covers the back of the holder when the plates are being exposed.

Figure 6 is a rear and side view of the circuit maker, conventionally called the exposing motor.

The motive power is an adjustable weight attached to a cord which is wound around a drum. Twenty-four binding posts are attached to the table at

the back of the exposing motor; other binding posts
are arranged for return or other currents.

Figure 7 illustrates a front and side view of the
upper part of the exposing motor. Fastened to the
frame is a ring of hard rubber, in which are inserted
twenty-four insulated segments of platinum-coated

Fig. 6.

brass; these segments are connected by insulated wires
to the twenty-four binding posts on the back of the
motor table, figure 6.

A shaft, connected by an arrangement of geared
wheels to the drum, passes through the center of the
segmented ring and carries a loose collar; a stout metal

rod is firmly attached near its longitudinal center to this loose collar. One arm of the rod carries a laminated metal scraper, or contact brush, arranged to travel around the periphery of the ring, and in its revolution to make contact with each segment in succession. The contact brush is connected through the arm with one pole

Fig. 7.

of the battery; and each segment— through its independent wire and magnet of the electro - exposors — with the other pole.

When twenty-four consecutive phases of an act of motion are to be photographed from one point of view, all of the insulated segments in the ring are put in circuit.

Fig. 7.

When twelve consecutive phases

are to be photographed synchronously from each of three points of view, each alternate segment is placed in circuit with the electric battery.

The manner in which the series of synchronous exposures is effected will be readily understood by reference to the diagram, 8.

ANIMAL LOCOMOTION.
DIAGRAM·OF·ELECTRICAL·CONNECTIONS·FOR·MAKING·CONSECUTIVE·PHOTOGRAPHIC
EXPOSURES·SYNCHRONOUSLY·FROM·SEVERAL·POINTS·OF·VIEW.

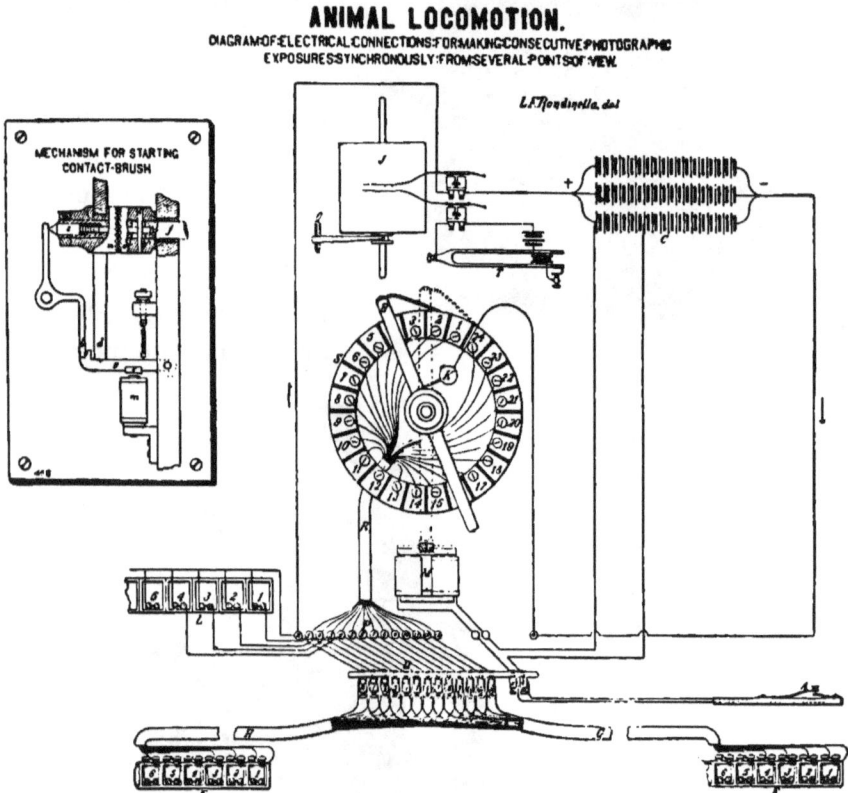

Fig. 8.

All being in readiness, and the weights and fan wheel adjusted to cause the contact brush to sweep over the periphery of the ring at the required rate of speed, the drum, and with it the shaft is set in motion.

At the proper time, pressure on a button completes an independent circuit through the magnet seen below the segmented ring, figure 7, and in the side diagram of figure 8.

The action of the armature releases the lower end of the rod on the loose collar, which, by means of a coiled spring, is immediately thrown into gearing with the already revolving shaft; the contact brush sweeps around the segmented ring and effects the consecutive series of exposures at the pre-arranged intervals of time.

At the University the intervals varied from the one-sixtieth part of a second to several seconds.

A record of these time intervals was kept by a chronograph, a well known instrument; it comprises a revolving drum carrying a cylinder of smoke-blackened paper, on which, by means of successive electric contacts, a pencil is caused to record the vibrations of a tuning fork, while a second pencil marks the commencement of each photographic exposure. The number of vibrations occurring between any two successive exposures marks the time. The tuning fork used made one hundred single vibrations in a second of time. To ensure greater minuteness and accuracy in the record, the vibrations were divided into tenths, and the intervals calculated in thousandths of a second.

For the purpose of determining the synchronous action of the electro-exposors while making a double series of exposures, the accuracy of the time intervals as recorded by the chronograph, and the duration of the shortest photographic exposures used in the investigation, the two batteries of portable cameras were placed side by side, and the exposors were each connected with the exposing motor by separate lengths of a hundred feet of cable.. The two series of cameras were pointed to a rapidly revolving disc of five feet diameter. The surface of the disc was black, with narrow white lines radiating from the center to the edge like the spokes of a wheel. A microscopic examination of the two series of resulting negatives proved that no variation could be discovered in the sychronous action of ten of the duplicated series of exposures, and that in the remaining two a variation existed in the simultaneity of a few ten-thousandths of a second — a result sufficiently near to simultaneity for all ordinary photographic work.

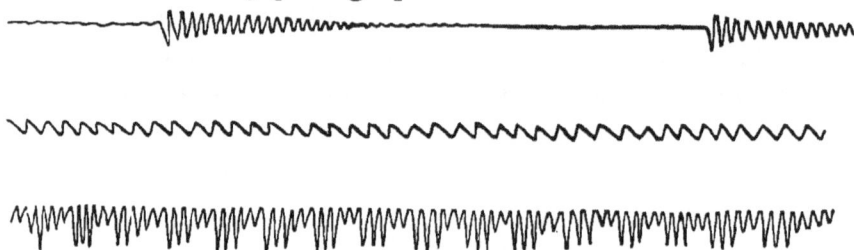

Fig. 9.

A reproduction of the chronographic record of one of these experiments is seen in figure 9.

The first line records the revolution of the disc; the second the vibration of the tuning fork: and each group of three long double markings in the third line indicates a photographic exposure.

The shortest exposures made at the University were — approximately — the one six-thousandth part of a second; such brief exposures are however for this class of investigation very rarely needed.

Some horses galloping at full speed will, for a short distance, cover about fifty-six or fifty-eight feet

of ground in a second of time; a full mile averaging
perhaps a hundred seconds. At this speed, a foot re-
covering its loss of motion will be thrust forward with
an occasional velocity of at least 120 lineal feet in a
second of time.

During the one one-thousandth part of a second the
body of the horse will at this rate move forward about
seven one-tenths of an inch, and a moving foot perhaps
one and a half inches, not a very serious matter for the
usual requirements of the amateur photographer.

A knowledge of the duration of the exposures,
however, was in this investigation of no value, and
scarcely a matter of curiosity, the aim always being to
give as long an exposure as the rapidity of the action
would permit, with a due regard to the necessary
sharpness of outline, and essential distinctness of detail.

The power used for operating the magnets, through
the exposing motor, was given from a lé Clanché
battery of fifty-four cells, arranged in multiple arc of
three series, each of eighteen cells.

During the investigation at the University of Pennsylvania, more than a hundred thousand photographic
exposures were made.

The negative plates were supplied by the Cramer
Dry Plate Company of St. Louis, and the positive
plates by the Carbutt Company of Philadelphia. On
a favorable day five hundred or six hundred negatives
were sometimes exposed; on one day the number of
exposures reached seven hundred and fifty.

The electrical manipulations were directed by Lino
F. Rondinella; the development room was in charge of
Henry Bell. The author takes pleasure in acknowl-

edging the skill, patience and energy which these gentlemen exhibited in their respective fields of labor.

Although the one six-thousandth part of a second
was the duration of the most rapid exposure made in
this investigation, it is by no means the limit of
mechanically effected photographic exposures, nor does
the one-sixtieth part of a second approach the limit
of time intervals. Marey, in his remarkable physiological investigations, has recently made successive exposures with far less intervals of time; and the author has
devised, and when a relaxation of the demands upon
his time permit, will use an apparatus which will
photograph twenty consecutive phases of a single
vibration of the wing of an insect; even assuming as
correct a quotation from *Nicholson's Journal* by Pettigrew in his work on Animal Locomotion that a common house fly will make during flight seven hundred
and fifty vibrations of its wings in a second of time, a
number probably far in excess of the reality.

The ingenious gentlemen who are persistently en-
deavoring to overcome the obstacles in the construction of an apparatus for aerial navigation, will perhaps
some day be awakened by the fact that the only suc-
cessful method of propulsion will be found in the action
of the wing of an insect.

We will now resume the subject proper of this
monograph.

It is impossible within its limits to trace the history
of the art of delineating animals in motion, or to
illustrate it with examples of the truthful impressions of
the primitive Artists, or of the imaginative and erro-
neous conceptions of many of those of modern times.

Certain phases of the facts of Animal Locomotion will alone be treated upon, as demonstrated by photographic research.

The illustrations and condensed definitions of the various gaits were prepared by the Author for the "Standard Dictionary." Before studying these it is essential that the meaning of the terms *step* and *stride* should be distinctly understood.

A STEP is an act of progressive animal motion, in which one of the supporting members of the body is thrust in the direction of the motion and the support transferred, wholly, or in part, from one member to another.

A STRIDE is an act of progressive animal motion, which, for its completion, requires all of the supporting members of the body, in the exercise of their proper functions, to be consecutively and regularly thrust in the direction of the movement until they hold the same relative positions in respect to each other as they did at the commencement of the notation. In the bipedal walk or run a step is one-half of a stride or full round movement. With all quadrupeds, except the kangaroo and other jumpers, *four* steps are necessary to complete the stride.

THE WALK.

The WALK is a method of progressive motion with a regular individual succession of limb movements. In the evolution of the terrestrial vertebrates the walk was probably the first adopted method of locomotion, and its execution is regulated by the law that the movement of the *superior* limb precedes the movement of its lateral *inferior* limb. This is proved not merely by the *ordinary* quadrupedal walk, but by the suspended motion of the sloth; the crawling of the child upon the ground, the erect walk of man; and the inverse limb movements of the ape tribe.

The relative time intervals of the foot-fallings vary greatly with many species of animals, and even with the same animal under different conditions.

Selecting the horse for the purpose of illustration we find that during the walk—his slowest progressive movement—he has always two, and for a varying period of time, or distance, three feet on the ground at once, while during a very slow walk the support will devolve alternately upon three feet and upon four feet.

If the notation of the foot-fallings commences with the landing of the right hind foot, the order in

SOME CONSECUTIVE PHASES OF THE WALK.

which the other feet are placed upon the ground will
be: the right fore, the left hind, and the left fore,
commencing again with the right hind.

Assuming that our observation of the stride of a
horse during an ordinary walk commences with the
landing of the right hind foot, the body will then be
supported by both hind and the left fore feet. The
left hind is now lifted, the support of the body devolves upon the diagonals—the right hind and left
fore—and continues so supported until the left hind
is in the act of passing to the front of the right; when
the right fore is next placed on the ground. The left
fore is now raised, and the body is supported by the
right laterals, until the landing of the left hind foot
relieves its fellow hind of a portion of its weight.
Two steps or one-half of a stride have now been made,
and with the substitution of the right feet for the left,
two other steps will be executed in practically the
same manner, and a full stride will have been completed. We thus see that during the walk a quadruped
is supported by eight different methods, the supporting limbs being consecutively:

Both hind and left fore.

Right hind and left fore *diagonals*.

Right hind and both fore.

Right hind and right fore *laterals*.

Both hind and right fore.

Left hind and right fore *diagonals*.

Left hind and both fore.

Left hind and left fore *laterals*.

Followed as at the commencement with both hind
and left fore.

When, therefore, during a walk, a horse is supported on two legs, with two feet suspended between them, each pair are laterals. On the other hand, when the suspended feet are respectively in advance of, and behind the supporting legs, each pair are diagonals.

These invariable rules have been unknown or ignored by many distinguished artists of modern times.

THE AMBLE.

The amble is a method of progressive motion with the same sequence of foot fallings as the walk, but in which a hind foot or a fore foot is lifted from the ground in advance of its fellow hind foot or its fellow fore foot being placed thereon. The support of the body therefore devolves alternately upon a single foot and upon two feet; the single foot being alternately a hind foot and a fore foot, and the two feet being alternately laterals and diagonals. At no time is the body entirely unsupported.

The following series of illustrations will clearly demonstrate the consecutive foot fallings and some characteristic phases of an ambling stride:

SOME CONSECUTIVE PHASES OF THE AMBLE.

The amble has various local names, such as the "single foot," the "fox trot," etc. It has sometimes been erroneously confused with the rack or the so-called "pace;" it is the most gentle and agreeable to the rider of all methods of locomotion of the horse, while the rack is the most ungraceful and disagreeable.

In Scott's romances are many allusions to the 'ambling palfry." Ben Jonson in "Every Man in His Humor" speaks of going "out of the old hackney-pace to a fine, easy amble," and Dickens in "Barnaby Rudge" refers to "the gray mare breaking from her sober amble into a gentle trot."

The ambling gait is natural to the elephant, and to the horse, the mule and the ass; but in many countries these latter animals are not encouraged in its use.

THE TROT.

The trot is a more or less rapid progressive motion of a quadruped in which the diagonal limbs act nearly simultaneously in being alternately lifted from and placed on the ground, and in which the body of the animal is entirely unsupported twice during each stride.

Selecting for the purpose of illustration the phases occurring during two steps or one-half of a stride of 18 feet in length by a horse trotting at the rate of a mile in two minutes and twelve seconds, we find that at the instant his right fore foot strikes the ground, the left hind foot is a few inches behind the point where it will presently strike. As the feet approach the ground, the right hind leg is drawn forward with the pastern nearly horizontal, while the left fore leg is flexed under the body. After the feet strike the ground and the legs approach a vertical position the pasterns are gradually lowered, and act as springs to break the force of the concussion until they are sometimes bent to a right angle with the legs.

At this period the fore foot is raised so high as to frequently strike the elbow, while the diagonal hind foot is comparatively but little above the ground, and is about to pass to the front of the left hind.

The pasterns gradually rise as the legs pass the vertical until the right fore foot has left the ground and the last propelling force is being exercised by the left hind foot; which accomplished, the animal is in mid air.

The right hind foot continues its onward motion

until it is sometimes much in advance of its lateral
fore foot, the former, however, being gradually lowered, while the latter is being raised. The right hind
and both fore legs are now much flexed, while the left
hind is stretched backwards to its greatest extent with
the bottom of the foot turned upwards, the left
fore leg is being thrust forwards and gradually

SOME CONSECUTIVE PHASES OF THE TROT.

straightened, with the toe raised as the foot approaches
the ground; which accomplished, with a substitution
of the left limbs for the right, we find them in the
same relative positions as when we commenced our
examination, and one-half of the stride is completed.

With slight and immaterial differences, such as
might be caused by irregularities of the ground, these

movements are repeated by the other pair of diagonals, and the stride is then complete.

If the stride of a trotting horse is divided into two portions, representing the comparative distances traversed by the aggregate of the body while the feet are in contact with, and while they are entirely clear of, the ground, the relative measurements will be found to vary very greatly, they being contingent upon length of limb, weight, speed, and other circumstances.

Heavily built horses will sometimes merely drag the feet just above the surface, but, in every instance of a trot, the *weight* of the body is really unsupported twice during each stride. It sometimes happens that a fast trotter, during the four steps of a stride, will have all his feet clear of the ground for a distance exceeding one-half of the length of the entire stride. Upon landing, a fore foot almost always precedes its diagonal hind.

It will be observed in the illustrations that while during the fast trot the fore feet are lifted so high that they frequently strike the breast, the hind feet are raised but little above the surface of the ground. The trot is common to all the single-toed and to nearly all the cloven-footed and soft-footed animals. It has, however, not been recorded as being adopted by the elephant, the camel, or the giraffe.

THE RACK.

The rack, sometimes miscalled the "pace," is a method of quadrupedal locomotion in which two lateral feet with nearly synchronous action are placed upon and lifted from the ground alternately with the other

laterals, the body of the animal being in the intervals
entirely without support. The distance which the
propelling feet hurl the animal through the air de-
pends, as with other movements, upon a variety of
circumstances; at a high rate of speed the distance will
be about one-half the total length of the stride. Upon

SOME CONSECUTIVE PHASES OF THE RACK.

landing, a hind foot usually precedes its lateral fore.

The rack is an ungraceful gait of the horse, and
disagreeable to those who seek comfort in riding.

The movements hitherto described are regular in
their action, and a stride may. be divided into two
parts, each of which — with a change of limbs — is
practically similar to the other; we now come to methods of progression which cannot be so divided, and
each stride must be considered as a unit of motion.

THE CANTER.

In the canter we discover the same sequence of foot fallings as in the walk, but not with the same harmonious intervals of time. The gait resembles the gallop in respect to its leaving the horse entirely unsupported for a varying period of time, and in the fact that the spring into the air is always effected from a fore foot, and the landing upon the diagonal hind foot; in other respects it materially differs from that method of progression.

Assuming that during a stride of the canter a horse springs into the air from a left fore foot, the right hind foot will first reach the ground; the two fore legs will at this time be flexed under the body, the right being the first landed, and for a brief period of time the support will devolve upon the laterals. The right fore foot is rapidly followed by the left hind. During a very slow canter the other fore foot will sometimes be landed in advance of the lifting of its diagonal, and the curious phase presented of all of the feet being in contact with the ground at the same instant. Usually, however, the first hind foot to touch the ground will be lifted, and the support thrown upon the diagonals.

The left fore is now brought down, and is followed by the lifting of the right fore; when the left laterals assume the duty of support. The left hind is now raised, and with a final thrust of the left fore foot the animal is projected into the air, to land again upon its diagonal, and repeat the same sequence of movements.

The above phases are selected from a single complete stride, in which the landing occurs on the *right* hind foot. Had the horse sprung from a *right* fore foot, the right and left feet would have been reversed through the entire series.

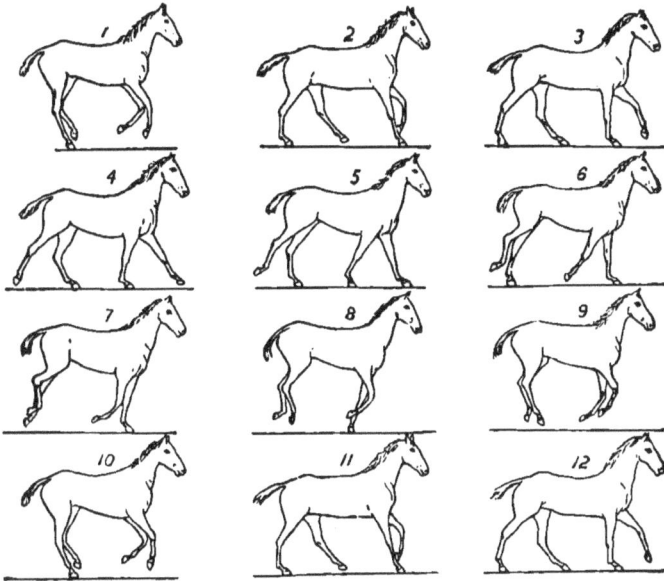

SOME CONSECUTIVE PHASES OF THE CANTER.

THE GALLOP.

The gallop is the most rapid method of quadrupedal motion; in its action the feet are independently brought to the ground; the spring into the air as in the canter is effected from a fore foot, and the landing upon the diagonal hind foot.

The phases illustrated are selected from the stride of a thorough-bred Kentucky horse, galloping at the rate of a mile in a hundred seconds, with a stride of about twenty-one lineal feet.

The length of stride and the distance which the

body is carried forward without support depend upon many circumstances, such as the breed, build and condition of the horse, speed, track, etc.

The phases illustrated and the measurement given apply to one stride of one horse, but may be considered as fairly representing the stride of a first-class horse in prime racing condition at the height of his speed, upon a good track.

Assuming—as in this instance—the springing into the air to have been effected from the right fore foot, the landing will take place in advance of the centre of gravity, upon the diagonal, or left hind foot; above, will be suspended the right hind foot, and at a higher elevation, several inches to the rear, will be the right fore foot, with the sole turned upward. The left fore leg will be in advance of the right, and also flexed. The force of the impact and the weight of the horse causes the pastern to form a right angle with the leg, and the heel is impressed into the ground.

The right hind foot strikes the ground and shares the weight of the body. The left hind foot leaves the ground while the right hind pastern is in its horizontal phase, supporting all the weight At this period the left fore leg is perfectly straight, with the toe much higher than the heel, and is thrust forward until the pastern joint is vertical with the nose, the right fore knee is bent at a right angle. The left fore foot now strikes and these diagonals are for a brief period upon the ground together. The left fore leg, however, immediately assumes the entire responsibility of support and attains a vertical position, with the pastern at a right angle. The right fore leg becomes perfectly rigid,

and is thrust forward to its fullest extent. The right
fore foot now strikes the ground, the two fore legs
form a right angle, and the hind feet are found thrust
backward, the right to its fullest extent. The left
fore leg having completed its functions of support, is
now lifted, and the weight transferred to the right fore

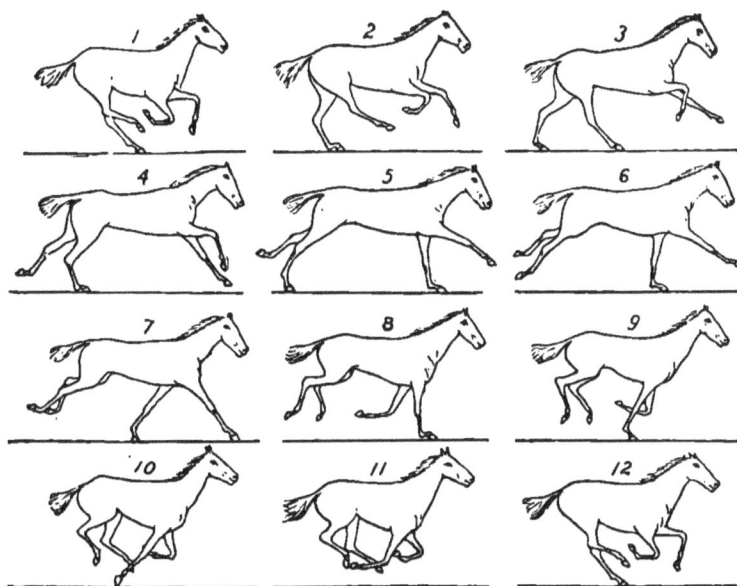

SOME CONSECUTIVE PHASES OF THE GALLOP.

foot alone, which is soon found behind the centre of
gravity; the left hind foot passes to the front of the
right fore leg, which, exercising its final act of propulsion, thrusts the horse through the air; the left
hind foot descends; the stride is completed, and the
consecutive phases are renewed. From this analysis
we learn that if the spring is made from the right fore

foot during the rapid gallop of. a thoroughbred horse,
it is supported consecutively by

The left hind foot.

Both hind feet.

The right hind foot.

The right hind and the left fore feet.

The left fore foot.

Both fore feet.

The right fore foot.

From which he springs into the air to re-commence
the phases with the left hind foot, while the only phase
in which he has been discovered without support is one
when the legs are flexed under the body. All of the feet
at this time are nearly close together and have comparatively little independent motion; this phase, therefore, more persistently than any other, forces itself
upon the attention of the careful observer, and conveys
to him the impression of a horse's rapid motion in
singular contradiction to the conventional interpretation, until quite recently, usually adopted by the
Artist.

It should not be understood that the term "spring"
implies that the body of the horse is greatly elevated
by that action; were it so, much force would be unneces-
sarily expended with the result of loss of speed. The
center of gravity of a horse trotting or galloping at a
high rate of speed will preserve an almost strictly
horizontal line, the undulations being very slight.

In the gallop of the horse it is probable there may
be sometimes a period of suspension between the lifting of one fore foot and the descent of the other, but
it has not yet been demonstrated.

The method of galloping described applies to the horse and its allies, and to most of the cloven and soft-footed animals.

In the gallop of the dog the sequence of foot falling and the action of the body is materially different, and the animal is free from support twice in each stride.

Assuming that a racing hound after a flight through the air with elongated body and extended legs (like the

THE GALLOP OF THE DOG.

conventional galloping horse), lands upon the left fore foot, the right fore will next touch the ground; from this he will again spring into the air, and with curved body and flexed legs land upon the right hind foot, while the right fore feet will be half the length of the body to the rear. The left hind now descends, another flight is effected, and again the left fore repeats its functions of support and propulsion.

These successive foot fallings are common to all dogs when galloping, and it is worthy of note that the same rotary action in the use of the limbs is adopted in the gallop of the elk, the deer and the antelope, all of which animals, like the dog, can for a time excel the horse in speed.

A search through all the dictionaries published at the time of writing, and accessible to the Author, fails to discover a correct definition of "the gallop." This motion is in America frequently miscalled the "run," and its execution "running," but no corresponding explanation of the word is given by any lexicographer.

In Scott's "Lady of the Lake" occurs "Then faint afar are heard the feet of rushing steeds in *gallop* fleet," many other distinguished Authors refer to the same action by the same name, by which, or its equivalents, it is universally known in Europe.

THE LEAP.

There is little essential difference in general characteristics of either of the several movements that have been described, but with a number of experiments made with horses while leaping, no two were found to agree in the manner of execution. The leap of the same horse at the same rate of speed, with the same rider, over the same hurdle, disclosed much variation in the rise, clearance, and descent of the animal. A few phases were, however, invariable. While the horse was raising his body to clear the hurdle, one hind foot was always in advance of the other, which exercised its last energy alone.

On the descent, the concussion was always first received by one fore foot, followed more or less rapidly by the other, sometimes as much as 30 inches in advance of where the first one struck; the hind feet were also landed with intervals of time and distance.

No attempt will be made to analyze the consecutive phases of various other acts of Animal Locomotion, such as rearing, bucking, kicking, tossing, etc., on account of the irregularity which characterizes their execution, and the difficulty of obtaining reliable data.

The Author has vainly sought for the rules which govern the hind feet of a playfully disposed mule; but the inquiry has usually been unsatisfactory, and upon some occasions disastrous. Should these movements be controlled by any general law, it is of such a complex nature that all attempts to expound it have hitherto been fruitless.

The figures in the series of circles (see appendix A) were selected from

"ANIMAL LOCOMOTION"

and arranged by the Author for his less ambitious work,

"POPULAR ZOOPRAXOGRAPHY."

(See Appendix C).

They were traced by the well known artist, Erwin Faber, and are reproduced one-third the diameter of the circles arranged for the zoöpraxiscope. Many of the original phases of movement are omitted on account of the optical law which in the construction of a zoöpraxiscope requires that the number of illustrations must bear a certain relationship to the number of perforations through which they are viewed.

The popular number of thirteen having been selected for the latter, the same number of figures illustrate actions without lateral progressive motion.

When the number of illustrated phases is less than the number of perforations, the succession of phases is in the direction of the motion, and the disc is necessarily revolved in a reverse direction.

When the number of phases is greater than the number of perforations, the phases succeed each other in a direction contrary to that of the motion, and the disc is revolved in the direction of the motion.

An increased or diminished number of figures will respectively result in an increased or diminished apparent speed of the object.

For further information on the subject, the reader is referred to the

ZOOPRAXISCOPE.

SYLLABUS OF A COURSE OF TWO LECTURES

ON

ZOOPRAXOGRAPHY

OR

THE SCIENCE OF ANIMAL LOCOMOTION IN ITS RELATION TO DESIGN IN ART.

Origin of the Author's Investigations—Diagram of the Studio at the University of Pennsylvania where the Investigation was conducted—Batteries of Cameras, Electro-exposers, Contact-motor, Chronograph, and other apparatus used for photographing consecutive phases of animal movements—Method of obtaining successive exposures of moving objects synchronously from several different points of view—Normal Locomotion of Animals —Twelve consecutive phases of a single step of the Horse while walking; also of the Ox, Elk, Goat, Buffalo, and other cloven-footed animals; the Lion, Elephant, Camel, Dog, and other soft-footed animals; of the Sloth while suspended by its claws, and of the Child while crawling on the ground; of man walking erect—The Normal Method of Locomotion by all animals essentially the same—The Quadrupedal Walk as interpreted by Prehistoric Man, by the Egyptians, Assyrians, Phœnicians, Etruscans, Greeks, Romans, Byzantines, and by eminent artists of mediæval and of modern times—The Statue of Marcus Aurelius the great source of modern errors; Marcus Aurelius in London, Edinburgh, Glasgow, Dublin, Paris, Berlin, Amsterdam, New York, Boston, and many

1

other cities—Albert Durer, Verrocchio, Meissonier, Paul Delaroche, Landseer, Rosa Bonheur, Elizabeth Thompson Butler, &c.—Other Quadrupedal movements, the Amble, Rack, Trot and Canter—Twelve phases in the Gallop of a Horse—Origin of the modern representation of the Gallop—Gallop as depicted by the Hittites, North American Indians, Egyptians, Assyrians, Greeks, the mediæval artists—The modern conventional gallop; evidences of its absurdity; acknowledgment by the Artist of the necessity of reformation—Leap of the Horse, Kick of the Mule, &c., all illustrated by photographs the size of life, from nature, and comparisons made with the interpretation of the same movements by artists of pre-historic, ancient, mediæval and modern times—Demonstration of the action of the primary feathers in the wing of a Bird while Flying, and a solution of the complex problem of Soaring.

AFTER THE VARIOUS METHODS OF LOCOMOTION HAVE BEEN DEMONSTRATED BY ANALYSIS, THEY WILL BE REPRESENTED SYN-THETICALLY BY THE ZOOPRAXISCOPE.

Among the many Institutions where Mr. Muybridge has had the honor of Lecturing on

ZOOPRAXOGRAPHY

are the following:—

Royal Academy of Arts, London.
Royal Society of London.
Royal College of Surgeons, London.
Royal Institution of Great Britain.
Royal Dublin Society.
Royal Geographical Society.
Royal Institution, Hull.
British Association for the Advancement of Science.
Linnean Society, Zoological Society.
Art and Science Schools, South Kensington Museum.
London Institution, Glasgow Philosophical Society.
Newcastle Literary and Philosophical Society.
Birmingham Natural History and Microscopical Society.
Town Hall, Birmingham; Nottingham Arts Society.
Manchester Athenæum.
University of Oxford.
Eton College, Clifton College.
Wellington College, Yorkshire College,
Rugby School, Charterhouse.
Leeds Mechanics' Institute.
Sheffield Literary and Philosophical Society.
Belfast Natural History and Philosophical Society.
Warrington Literary and Philosophical Society.
Yorkshire Philosophical Society, Bristol Naturalists' Society.
Bath Associated Scientific and Art Societies.
Ipswich Scientific Society, Photographic Society of Ireland.
Liverpool Associated Literary, Scientific and Art Societies.
St. George's Hall, Liverpool.
School of Military Engineering, Chatham.
The School of Fine Arts; Hall of the Hemicycle, Paris.
The Society of Artists, Berlin.
The Society of Artists, Vienna.
The Society of Artists, Munich.
The Urania Scientific Society, Berlin.
The Polytechnic High School, Vienna.
The Polytechnic High School, Munich.
The University of Turin.
The "Cercle de L'Union Artistique,"
The Studio of M. Meissonier in Paris, Etc., Etc., Etc.

And at all the principal Institutions of Art, Science, Education and Learning in the United States of America.

1 ATHLETE, HORSE-BACK SOMERSAULT.

ABBREVIATED CRITICISMS.

"On Monday last, in the theatre of the ROYAL INSTI-
TUTION, a select and representative audience assembled to
witness a series of the most interesting demonstrations of
Animal Locomotion given by Mr. Muybridge.

"The Prince and Princess of Wales, Princess Victo-
ria, Louise, and Maud, and the Duke of Edinburgh honored the occasion by their presence; likewise did I note
among the brilliant company Earl Stanhope, Sir Frederick
Leighton, P.R.A.; Professors Huxley, Gladstone, and

2. ATHLETES BOXING.

Tyndall; and last, not least, Lord Tennyson, poet lau-
reate.

"Mr. Muybridge exhibited a large number of photographs of horses galloping, leaping, etc. . . . By
the aid of an astonishing apparatus called a ZOOPRAXI-
SCOPE, which may be briefly described as a magic lantern
run mad (with method in the madness), the animals
walked, cantered, ambled, galloped, and leaped over hur-
dles in a perfectly natural and lifelike manner. I am
afraid that, had Muybridge exhibited his ZOOPRAXISCOPE
three hundred years ago, he would have been burned as a

3. ATHLETES RUNNING.

wizard. . . . After the horses came dogs, deer, and
wild bulls.   Finally man appeared (in instantaneous photography) on the scene, and ran, leaped, and turned back
somersaults to admiration."—GEORGE AUGUSTUS SALA in
Illustrated London News.

"Both scientific and artistic circles in London are at
present greatly interested in the triumphs of Mr. Eadweard
Muybridge in photographing the successive phases of animal movements.   Our leading biologists and artists have
at once perceived and acknowledged the vast importance
of the results of his work."—*The Times, London*.

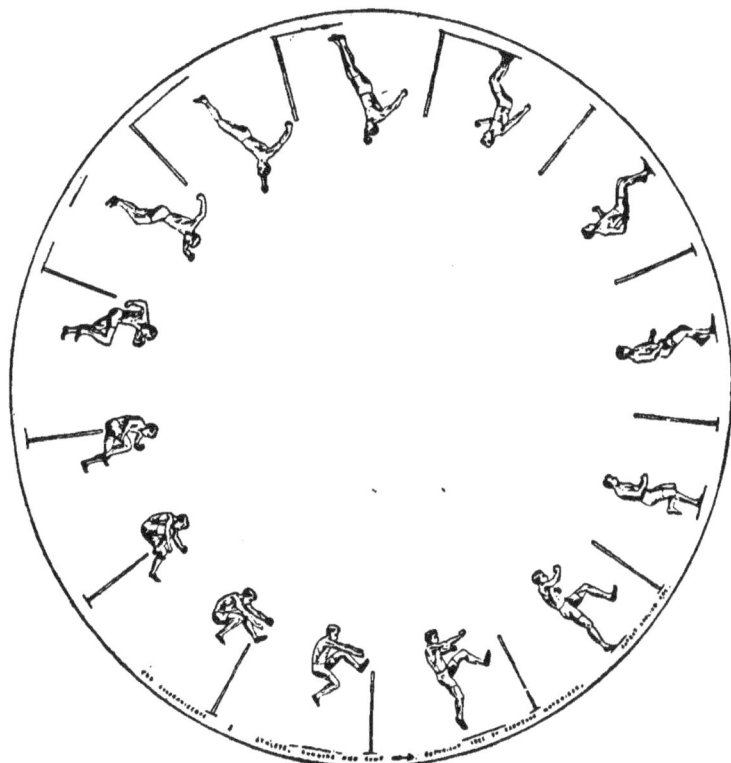

5. ATHLETE, RUNNING HIGH JUMP.

"The Archbishop of York occupied the chair. . . . His Grace congratulated the crowded and distinguished audience on the opportunity afforded them of hearing Mr. Muybridge, and said that to everybody who felt an interest in the phenomena of motion, the magnificent results of the investigation carried on by Mr. Muybridge and the University of Pennsylvania were wonderfully instructive." — *York Herald.*

"His audiences have been drawn from the very first ranks of art, science, and fashion."—*British Journal of Photography.*

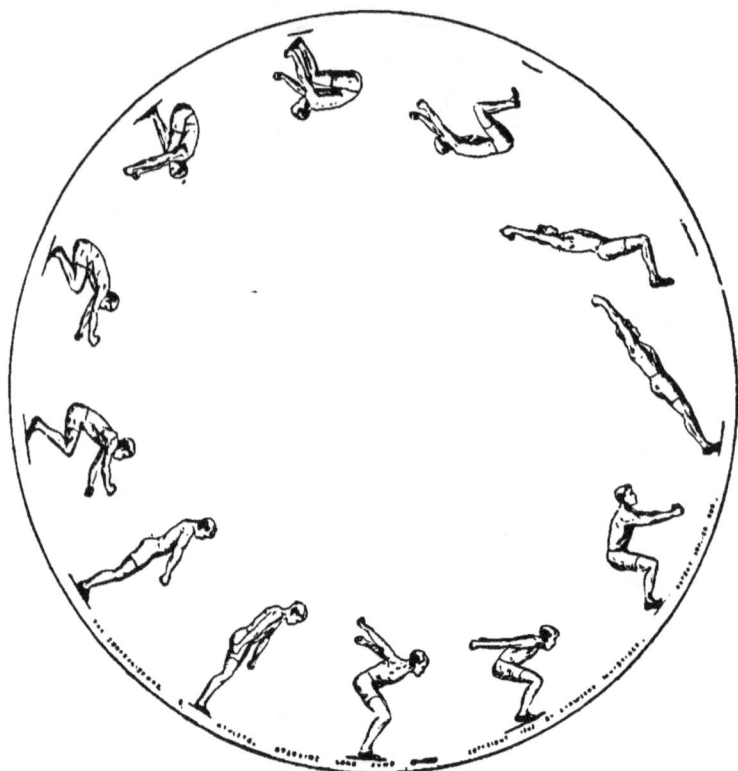

6. ATHLETE, STANDING LONG JUMP.

"These demonstrations are marvellously complete,
. . . exceedingly abundant and rich in suggestion and
instruction, and appeal to almost every class or condition
of humanity."—*Saturday Review, London.*

"Mr. Muybridge delighted his audience with his wonderful photographs."—*The Times, London.*

" . . . Last night Mr. Muybridge gave his final
lecture in Newcastle on 'The Science of Animal Locomotion,' with the whole of the wonderful illustrations; the
Art Gallery being again crowded to excess."—*Newcastle
Chronicle.*

11. ATHLETES. BASE BALL; BATTING.

" A photographic achievement which seemed to me at the time scarce credible, and which I was presently assured by one of our ablest English photographers was absolutely outside the bounds of possibility."—PROFESSOR R. A. PROCTOR in the *Gentleman's Magazine.*

"At the conversazione of the Royal Society much interest was excited by Mr. Eadweard Muybridge's lecture. The ZOOPRAXISCOPE afforded the spectator an opportunity of studying by synthesis, the facts of motion which are also demonstrated by analysis."—*Illustrated London News.*

14. Boys Playing Leap-frog.

"A really marvellous series of plates."— *Nature, London.*

"Artistic people are all talking about Mr. Muybridge, who has come hither with that rare desideratum—*something new.*"—London Correspondence, *Philadelphia Times.*

"It is impossible to do justice in this short time to the extraordinary exhibition given by Mr. Muybridge at the Institute of Technology. . . . The interest they excite in the mind of the spectator is indescribable."—*Sunday Gazette, Boston.*

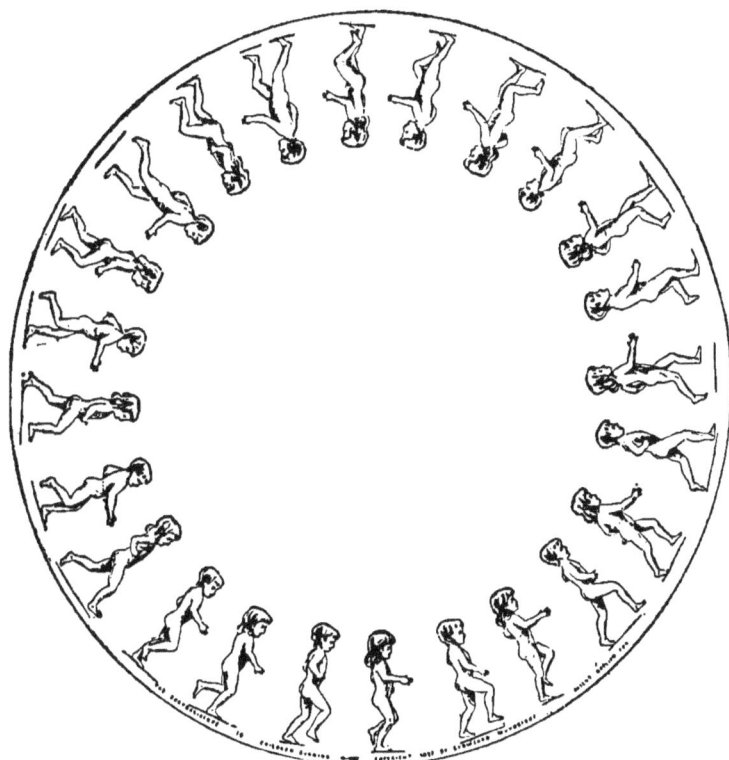

16. CHILDREN RUNNING.

"The photographs have solved many complicated questions as to animal locomotion."—*Art Journal, London.*

"The effect was weird, yet fascinating. Plaudit followed plaudit. A better pleased assemblage of people it would be difficult to find."—*Boston Journal.*

" . . . Mr. Muybridge then gave his famous lecture and demonstration on Animal Locomotion. The hall (St. James') was crowded, and many were unable to obtain seats."—Report of the Photographic Convention, *British Journal of Photography.*

17. Elephant Ambling.

"A demonstration that vividly interests all the world."
—*L'Illustration, Paris.*

"Many of these pictures have great—indeed, astonishing—beauty. The interest which they present from the scientific point of view is three-fold :—(*a*) They are important as examples of a very nearly perfect method of investigation by photographic and electrical appliances. (*b*) They have also a great value on account of the actual facts of natural history and physiology which they record. (*c*) They have, thirdly, a quite distinct, and perhaps their most definite, interest in their relation to psychology."—
Prof. E. Ray Lankester, F. R. S., in *Nature.*

18. Lion Walking

" Mr. Meissonier's critical guests were evidently scep-
tical as to the accuracy of many of the positions; but when
the photographs were turned rapidly, and made to pass
before the lantern, their truthfulness was demonstrated
most successfully."—*Standard*, *London*.

"Meissonier, devoting himself to his friends, evidently cared little for personal compliments; he was anxious
for the well-deserved distinction of his *protégé* Muybridge.
.   .   .   'C'est merveilleusement arrangé!' said Alex-
andre Dumas. 'C'est que la nature *compose* crânement
bien!' replied Meissonier."—*Le Temps*, Paris.

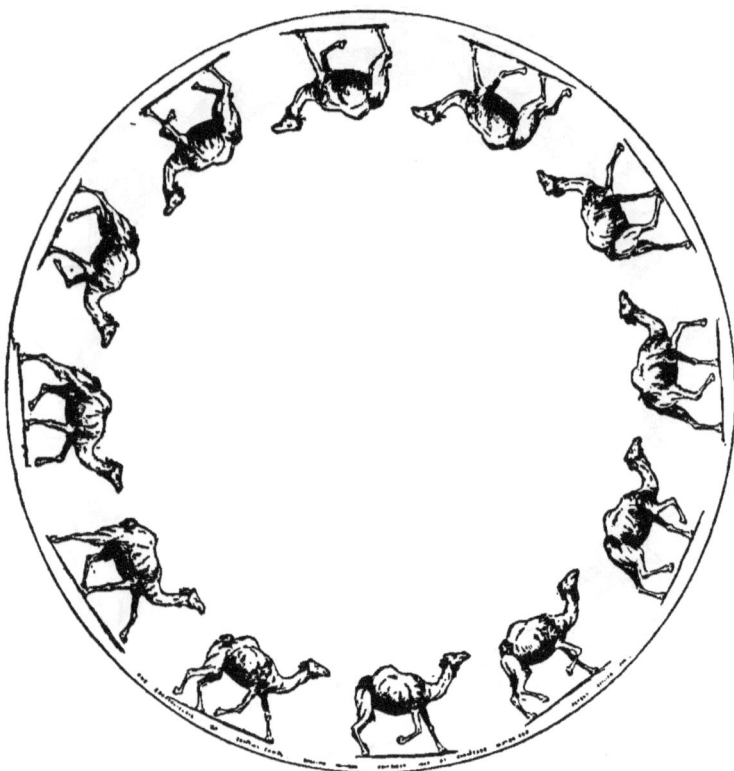

20. EGYPTIAN CAMEL RACKING.

"The sensation of the day, and the topic of popular conversation."—*Boston Daily Advertiser.*

"The rapid movements by different animals were most interesting: and hurdle-racing by horses—the very whipping process being visible—brought down the house."—*Boston Herald.*

"On revolving the instrument, the figures that have been derided by so many as impossible absurdities, started into life, and such a perfect representation of a racehorse at full speed as was never before witnessed was immediately visible."—*The Field, London.*

21. BABOON WALKING.

"Mr. Muybridge showed that many of our best artists
have been in the habit of depicting animals in positions
which they never assume in nature."—*Chambers' Edinburgh Journal.*

"The large school-room (Clifton College) was crowded.
The head master presided. Loud applause and frequent
laughter greeted the life-sized photographs from nature,
which by a rapid revolution of the ZOOPRAXISCOPE, showed
among other actions, the ambling of an elephant, the
gallop of a race-horse, the somersault of a gymnast and
the flight of a bird."—*Bristol Mercury.*

22. KANGAROO JUMPING.

"The lecture theatre of the ROYAL ACADEMY OF ARTS was filled to overflowing."—*Athenæum, London.*

"The Royal Dublin Society's Theatre was filled to its utmost capacity yesterday afternoon, when Mr. Muybridge resumed his course of Lectures. The demonstration is simply marvellous."—*Daily Express, Dublin.*

"The result of years of labor, and of large expenditure of money is at last laid before the public in this magnificent work, and the result is one of which Mr. Muybridge and the University of Pennsylvania may well be proud."—*Evening Post, New York.*

23. BUFFALO GALLOPING.

" A Lecture of an exceptionally interesting character."
—*Nottingham Guardian.*

"There was a crowded attendance. Throughout the lecture Mr. Muybridge retained the close interest of his audience, and drew from them frequent and warm applause."—*The Scotsman, Edinburgh.*

" In all my long experience of London life I cannot recall a single instance where such warm tributes of admiration have been so unsparingly given by the greatest in the land, as in the case of Mr. Muybridge's lectures."— OLIVE LOGAN in the *Morning Call, San Francisco.*

24. ELK GALLOPING.

"Mr. Muybridge illustrated his lecture with a series of most valuable photographs, as well as that most fascinating of scientific toys—the ZOOPRAXISCOPE."—*Magazine of Art, London.*

"His labors attracted considerable attention in the world of science, while among artists and art critics a pretty controversy set in on the subject of the horse and his representation in art, which is likely to be revived and extended to other fields. . . . With Mr. Muybridge, 'Instantaneous Photography' has acquired a new significance. . . ."—*Saturday Review, London.*

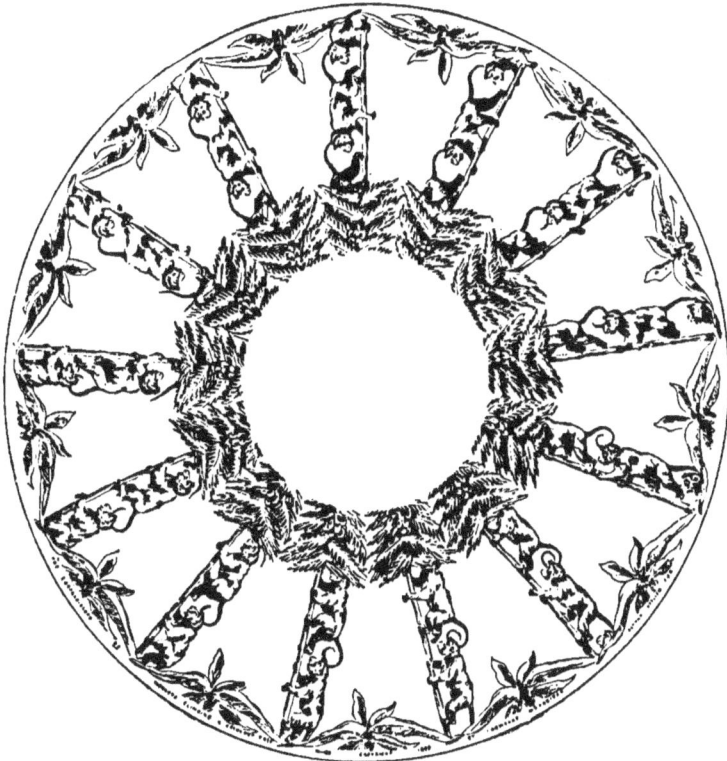

25. MONKEYS CLIMBING A COCOA PALM.

"No parallel in the history of photography."--*Photographic Times, New York.*

"An exhibition which Raphael, Tintoretto, Michael Angelo, and other great masters of the Renaissance would have travelled all over Europe to see."—*Evening Transcript, Boston.*

"The audience was astonished and delighted at the marvellous demonstrations of Animal Locomotion that were brought before them. . . . The most remarkable feature of the British Association meeting this year." —*Newcastle Journal.*

28. Greyhound Galloping.

" The effects of the Zoopraxiscope made up one of the most unique and instructive entertainments imaginable."—*Boston Daily Globe.*

"A more curious, entertaining, and suggestive exhibition it has not been our good fortune for a long time to attend."—*Sacramento Record-Union.*

"Everybody has heard something of the wonderful success which Mr. Muybridge has achieved; and in no country in the world is greater interest felt in his work, particularly as regards horses, than in England."—*Engineering, London.*

29. Mule, Bucking and Kicking.

"Simply marvels of the photographer's art."—*Mercury*, Leeds.

"Not the least instructive part of the Lecture was the contrast between the positions of animals as shown in ancient and modern art, with their true positions as shown by themselves in the camera."—*New York Tribune.*

"Professor Marey invited to his residence a large number of the most eminent men in Europe for the purpose of meeting Mr. Muybridge, and witnessing an exhibition that should be placed before the whole Parisian public."—*Le Globe, Paris.*

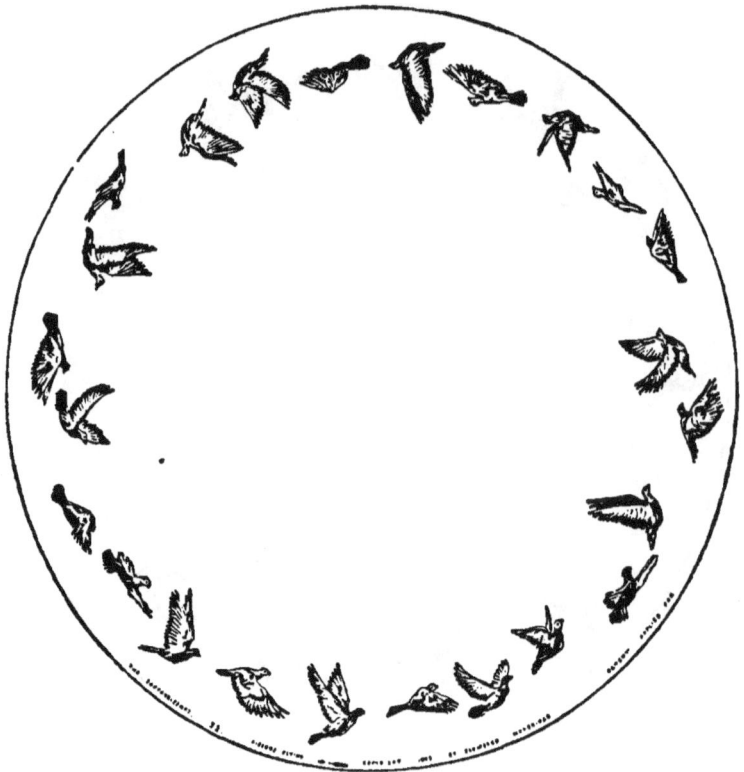

32. Pigeons Flying.

"The art critic and the connoisseur will find a study of Mr. Muybridge's work of inestimable value in aiding them to criticize intelligently."—*Pennsylvanian, Philadelphia.*

"The applause which greeted these wonderful pictures from the brilliant company was hearty in the extreme; and all predicted a new era was open to art, and new resources made available for the use of artists."—*Galignani's Messenger, Paris.*

"Of immense interest and value."—*Lippincott's Magazine, Philadelphia.*

34. GRECIAN DANCING GIRLS

"The ZOOPRAXISCOPE is the latest, most unique, and instructive form of amusement possible."—*Commercial Gazette*, Cincinnati.

"His work at once attracted the attention of the world."—*Scientific American*, New York.

"Of much interest and value, as well as a source of great amusement."—*Observer*, London.

"The realism of the motions of the various animals was intense, and the audience was very enthusiastic."—*Boston Post*.

39. HORSE TROTTING (fast).

"The Lecturer proceeded to show enlarged photographs of various animals in motion, as the horse, dog, lion, mule, cat, etc. . . . These were followed by some very striking pictures of the flight of birds, which from a scientific standpoint were by far the most interesting and valuable of the photographs shown during the evening."—*Lancet*, London.

"Of extreme interest, not only to the artists and scientists, but to the greater part of his audience, who were neither the one or the other."—*Birmingham Daily Gazette*.

41. Horse Cantering.

"A host of well-known scientists and artists are greatly interested in this remarkable work."—*Pall Mall Gazette.*

"The lecture on Tuesday night more than fulfilled the expectations which the audience had formed of Mr. Muybridge's researches."—*Belfast News Letter.*

"Mr. Muybridge might well be proud of the reception accorded him by his distinguished audience; it would have been difficult to add to the *éclat* of his appearance, and his lecture was welcomed by a warmth as hearty as it was spontaneous."—*The Photographic News, London.*

42. Horse Galloping

"The illustrations are truly wonderful, and the rapid changing positions were most instructive."—*Nottingham Express.*

"The concert room was crowded. . . . A vote of thanks to the Lecturer was proposed by his Grace the Archbishop."—*Yorkshire Chronicle.*

"A very brilliant audience was assembled at the Royal Institution. . . . The photographs properly studied should be most valuable in affording truer and more exact data for the painter to base his work upon. . . ."—*The Builder, London.*

43. Horse Jumping.

"A very important subject to all those interested in art."—*Belfast News Letter*.

"It is now nine years since the photographs of Mr. Eadweard Muybridge surprised the world by challenging all received conceptions of animal motion."—*Century Magazine, New York*.

"The interest excited by the novelty, both of the demonstrations and the results, was so great, that Mr. Muybridge has been invited by the Photographic Society of Ireland to repeat them to-night in a public lecture."— *The Freeman's Journal, Dublin*.

44. HORSE HAULING.

" The audience filled the large hall, and by their frequent and hearty applause, expressed their appreciation of
the lecture."—*Irish Times, Dublin.*

" A very large audience again assembled in the Town
Hall last evening, on the occasion of the second Lecture
by Mr. Muybridge. The Mayor, who presided, referred
to the first Lecture as perhaps the most unique ever deliv-
ered in Birmingham."—*Birmingham Daily Gazette.*

" The attendance was exceedingly large, and the Lecture and admirable illustrations were loudly applauded."
—*The Irish Times, Dublin.*

45. COLUMBIAN EXPOSITION HORSE RACE, GALLOPING.

"There was a very large attendance, and seldom have we seen so much genuine admiration and enthusiasm displayed as were evoked by Mr. Muybridge's illustrations, which were really wonderful." — *The Daily Express, Dublin.*

"There was a crowded audience, and the Lecture, which was listened to with the greatest interest, was warmly applauded."—*The Freeman's Journal, Dublin.*

"No description can do justice to the extent and variety of the subjects presented in this thorough study of animal movements."—*Ledger,* Philadelphia.

46. COLUMBIAN EXPOSITION HORSE RACE, TROTTING

"Wonderful and interesting demonstration; its influence will become more and more potent and universal as the years go on."— *Argus, Albany.*

"Will necessarily revolutionize the treatment of the action of the horse in painting and sculpture. For the physiological study of animal movements these pictures are a veritable treasure."—*Landwirthschaftliche-Zeitung, Vienna.*

"I am lost with admiration of these photographs of Mr. Muybridge."—PROFESSOR MAREY, in *La Nature, Paris.*

47. COLUMBIAN EXPOSITION SPEEDWAY.

" Interesting and instructive to all."—*New York Herald.*

" Highly interesting and valuable for every lover of horses."—*Illustrirte Zeitung, Berlin.*

" We cannot more fittingly conclude our review than by repeating our recommendation of the work to all artistic and scientific bodies."—*The Nation, New York.*

" So perfect was the synthesis that a dog in the lecture room barked and endeavored to chase the phantom horses as they galloped across the screen." —*Berkeley Weekly News.*

48. VILLAGE BLACKSMITHS.

" Noted artists, such as Menzel, Knaus, Begas; eminent
scientists, such as von Helmholtz, Siemens and Förster
and even the imperturbable field-marshal, Count von
Moltke, were enthusiastic in their applause."—*Illustrirte
Zeitung*.

" A very large number could not obtain admission, so
great was the desire to hear the lecture. . . . A wonderful surprise even to the careful observer of Nature."—
Die Press, Vienna.

" The lecture was received with stormy applause."—
Berliner Post, Berlin

' The lecture was given in a popular manner, with

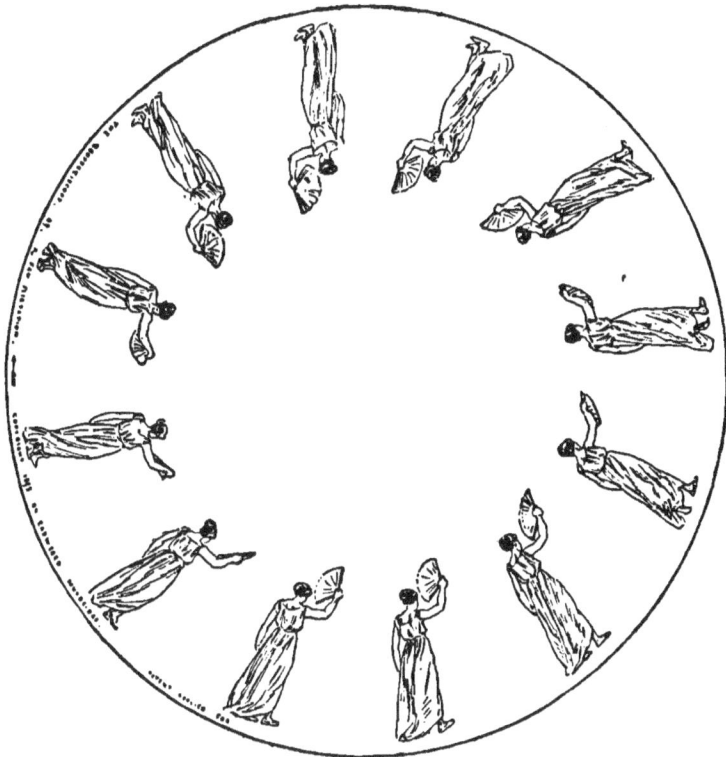

49. A FAN FLIRTATION.

scientific accuracy and artistic taste. . . . . The room was filled to the last corner; nearly all the Royal Family and the Ministers were present."—*Münchener Neueste Nachrichten*, Munich.

"After attending Mr. Muybridge's demonstrations, we felt no surprise at his having been received so enthusiastically in Paris."—*Berliner Tageblatt*, Berlin.

"The lectures of Mr. Muybridge are unquestionably the most intensely interesting we ever listened to. No one in Berlin should fail to attend them."—*Norddeutsch Allgem Zeitung*, Berlin.

"Some lectures are too technical for the general public.

50. ATHLETE, RUNNING LONG JUMP.

Here is one in which everybody is interested. The Lecture Theatre was crammed to repletion; we thought a few vacant places might have been reserved for those whose pleasant duty it is to record the brilliant success of Mr. Muybridge."—*Pall Mall Budget*, London.

"So great an interest did the demonstrations excite that Mr. Muybridge was unanimously requested to repeat them. Two days afterward this distinguished company, including the venerable Field-Marshal (Count von Moltke) himself, attended a repetition of the lecture."—*Illustrirte Zeitung*.

APPENDIX B.

ANIMAL LOCOMOTION.

DESCRIPTION OF THE PLATES.

The results of the investigation executed for the University of Pennsylvania are

SEVEN HUNDRED AND EIGHTY-ONE SHEETS OF ILLUS-
TRATIONS,

containing more than 20,000 figures of men, women, and children, animals and birds, actively engaged in walking, galloping, flying, working, jumping, fighting, dancing, playing at base-ball, cricket, and other athletic games, or other actions incidental to every-day life, which illustrate motion or the play of muscles.

These sheets of illustrations are conventionally called "plates."

EACH PLATE IS COMPLETE IN ITSELF WITHOUT REFERENCE
TO ANY OTHER PLATE,

and illustrates the successive phases of a single action, photographed with automatic electro-photographic apparatus at regulated and accurately recorded intervals of time, *consecutively* from one point of view; or, *consecutively* AND *synchronously* from *two*, or from *three* points of view.

A series of twelve consecutive exposures, from each of the three points of view, are represented by an outline tracing on a small scale of plate 579, a complete stride of a horse walking; the intervals of exposures are recorded as being one hundred and twenty-six one-thousandths of a second.

1

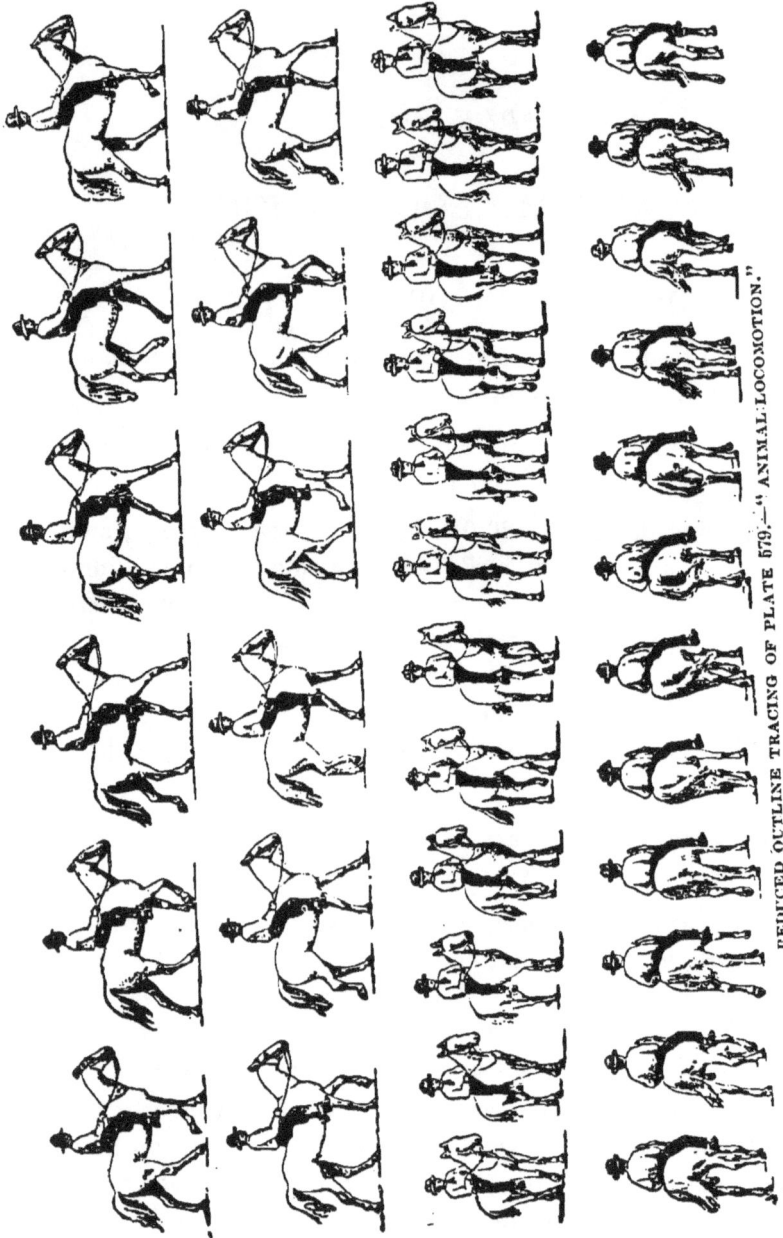

REDUCED TRACING OF SOME PHASES FROM PLATE 759.

REDUCED TRACINGS OF PLATE 347.

When one of the series of foreshortenings is made at a right angle with the lateral series the arrangement of the phases is usually thus:

												Laterals.
1	2	3	4	5	6	7	8	9	10	11	12	Re a r Foreshortenings from points of view on the same vertical line, at an angle of 90 deg.
1	2	3	4	5	6	7	8	9	10	11	12	from the Laterals.
1	2	3	4	5	6	7	8	9	10	11	12	Front Foreshortenings from points of view on the same horizontal plane, at suitable angles from the Laterals.

The plates are not *photographs* in the common acceptation of the word, but are printed in PERMANENT INK, from gelatinised copper-plates, by the New York Photo-Gravure Company, on thick linen plate-paper.

The size of the paper is 45 × 60 centimetres—(19 × 24 inches), and the printed surface varies from 15 × 45 to 20 × 30 centimetres—(6 × 18 to 9 × 12 inches).

The number of figures on each plate varies from 12 to 36.

To publish so great a number of plates as one undivided work was considered unnecessary, for each subject tells its own story; and inexpedient, for it would defeat the object which the University had in view, and limit its acquisition to wealthy individuals, large Libraries, or Institutions where it would be beyond the reach of many who might desire to study it.

It has, therefore, been decided to issue a series of One Hundred Plates, which number, for the purposes of publication, will be considered as a "COPY" of the work. These one hundred plates will probably meet the requirements of the greater number of the subscribers.

In accordance with this view is re-issued the following prospectus.

.

PROSPECTUS

ANIMAL LOCOMOTION,

AN ELECTRO-PHOTOGRAPHIC INVESTIGATION OF CONSECUTIVE
PHASES OF ANIMAL MOVEMENTS,

BY

EADWEARD MUYBRIDGE.

Commenced, 1872—Completed, 1885.

PUBLISHED 1887, UNDER THE AUSPICES OF THE

UNIVERSITY OF PENNSYLVANIA.

Exclusively by Subscription.

CONSISTING OF A SERIES OF

ONE HUNDRED PLATES,

AT A SUBSCRIPTION PRICE OF

ONE HUNDRED DOLLARS

For the United States, or

TWENTY GUINEAS

For Great Britain;

Or the equivalent of Twenty Guineas in the gold currency
of other countries in Central or Western Europe.

The Plates are enclosed in a strong, canvas-lined, full
AMERICAN-RUSSIA LEATHER PORTFOLIO.

Additional Plates in any required number will be supplied to the subscriber at the same proportionate rate;
these, however, must be ordered at the same time as the
subscription Plates.

It was considered inadvisable to make an *arbitrary*
selection of the one hundred Plates offered to subscribers,
and with the object of meeting, as far as possible, their
diverse requirements, they are invited to make their own
selection, either from the subjoined list of subjects, **or**

from a detailed catalogue, which will be forwarded free
of expense to every subscriber.

The following are the numbers of Plates published of
each class of subjects, from which the subscriber's selection can be made:—

Class.			Plates Published.
1. Men,	draped		6
2. "	pelvis cloth		72
3. "	nude		133
4. Women, draped			60
5. "	transparent drapery and semi-nude		63
6. "	nude		180
7. Children, draped			1
8. "	nude		15
9. Movements of a man's hand			5
10. Abnormal movements, men and women, nude and semi-nude			27
11. Horses walking, trotting, galloping, jumping, &c.			95
12. Mules, oxen, dogs, cats, goats, and other domestic animals			40
13. Lions, elephants, buffaloes, camels, deer, and other wild animals			57
14. Pigeons, vultures, ostriches, eagles, cranes and other birds			27

Total number of Plates.............. 781
Containing more than 20,000 Figures.

Should the selection be made from the Catalogue, it
will be advisable to give the Author permission to change
any one of the selected Plates for any other illustrating
the same action, if, in his judgment, the substituted Plate
illustrates that action with a better model, or in a more
perfect manner than the one selected.

With regard to the selection of Plates, however, it
has been found by experience that unless any special subject or plate is required it will be more satisfactory to the
subscriber if he gives the Author GENERAL IN-

STRUCTIONS as to the CLASS of subjects desired and to leave the SPECIFIC selection to him.

Many of the large Libraries and Art or Science Institutions in America and in Europe have subscribed for, and have now in their possession, a complete series of the seven hundred and eighty-one Plates, the subscription price for which is

FIVE HUNDRED DOLLARS

in the United States,

ONE HUNDRED GUINEAS

in Great Britain for the complete series, in eight full AMERICAN-RUSSIA LEATHER PORTFOLIOS, or if bound in eleven volumes, each plate *hinged*, full American-Russia leather,

FIVE HUNDRED AND FIFTY DOLLARS

in the United States,

ONE HUNDRED AND TEN GUINEAS

in Great Britain; or its equivalent for any city in Central or Western Europe.

Subscribers who wish to make use of these Plates for the promotion or diffusion of knowledge, or for artistic or scientific purposes, will be afforded facilities for acquiring working copies by special arrangement with the Author.

The investigations of the Author are so well known; and so generally recognized as affording the only basis of truthful interpretation or accurate criticism of Animal Movement, that it is perhaps scarcely necessary to quote from the many elaborate reviews of "Animal Locomotion," which have been published in the American, English, French, and German Scientific, Artistic, and other Journals. A few extracts therefrom are however given in Appendix A.

For the value of the present work to the general stu-

dent of Nature and the lover of Art, no less than to the
Artist and the Archæologist, the Physiologist and the
Anatomist, it is with much pride and gratitude that he
refers to the annexed list of some of his subscribers.

SUBSCRIBERS.

The general or departmental Libraries of the following

UNIVERSITIES.

Amsterdam	Freiburg	Königsberg	Prag
Andrews, St.	Genève	Leiden	Roma
Basel	Genova	Leipzig	Rostock
Berlin	Glasgow	Liège	Strassburg
Bern	Göttingen	Louvain	Torino
Bologna	Griefswald	München	Tübingen
Bonn	Hallé	Napoli	Utrecht
Breslau	Heidelberg	Oxford	Wien
Bruxelles	Innsbrück	Padova	Würzberg
Edinburgh	Jena	Pisa	Zürich
Erlangen	Kiel		

IMPERIAL, NATIONAL, OR ROYAL ACADEMIES OF FINE ARTS.

Amsterdam	Budapest	Liège	Roma (de
Antwerpen	Dresden	London	France)
Berlin	Düsseldorf	Manchester	Sheffield
Bern	Firenze	Milano	Torino
Birmingham	Frankfurt	München	Venezia
Bologna	Genova	Napoli	Wien
Breslau	Gent	Paris	Zürich
Bruxelles	Leipzig	Praha	

Architectural Institute, München
Herkomer School of Art, Bushey

ART MUSEUMS.

Amsterdam	Berlin	Budapest

ARCHÆOLOGICAL INSTITUTES AND MUSEUMS,

Dresden	Königsberg	Rostock	Würzburg

| Griefswald | Leipzig | Strassburg | Zürich |
| Heidelberg | Prag | Wien | |

INDUSTRIAL ART AND SCIENCE MUSEUMS.

| Berlin | Edinburgh | Paris | Wien |
| Dublin | Kensington | | |

INDUSTRIAL ART SCHOOLS.

| Amsterdam | Budapest | Nürnberg | Zürich |
| Breslau | Frankfurt | | |

LIBRARIES.

The Royal Library, Windsor Castle.
Imperial Library, Berlin.

Birmingham, Free Public	London, British Museum
Edinburgh, Advocates'	Manchester, Free Public
Glasgow, Mitchell Free	Nottingham, Free Public
Liverpool, Free Public	Paris, National Library

ANATOMICAL INSTITUTES.

Bern	Innsbrück	München	Tübingen
Breslau	Kiel	Pisa	Würzburg
Freiburg	Königsberg	Prag	Zürich
Hallé	Leipzig	Rostock	

ROYAL COLLEGES OF SURGEONS.

Edinburgh   London

PHYSIOLOGICAL INSTITUTES.

Basel	Freiburg	Kiel	Strassburg
Berlin	Genova	Königsberg	Torino
Bern	Göttingen	Leipzig	Tübingen
Bologna	Griefswald	Louvain	Wien .
Bonn	Hallé	München	Würzburg
Breslau	Heidelberg	Napoli	Zürich
Bruxelles	Innsbrück	Prag	
Erlangen	Jena	Rostock	

VETERINARY INSTITUTES.

| Alfort | Bern | Berlin | Dresden | London |

ANTHROPOLOGICAL MUSEUMS.

| Dresden | Firenze |

ETHNOLOGICAL, NATURAL HISTORY, AND ZOÖLOGICAL INSTITUTES AND MUSEUMS.

Amsterdam	Kiel	Liège	Paris
Bruxelles	Leiden	Napoli	Rostock
Freiburg			

PHYSICAL INSTITUTES.

Basel	Genève	Prag	Rostock
Bologna	Heidelberg	Roma	Utrecht
Bruxelles	Padova		

POLYTECHNIC HIGH SCHOOLS.

| Berlin | Firenze | Wien | Zürich |

COLLEGES.

| Charterhouse | Clifton | Dublin (Trin.) |
| Eton | Owens | Wellington |

ROYAL PORCELAIN MANUFACTORIES.

| Berlin | Dresden |

ARTISTIC, LITERARY OR SCIENTIFIC CLUBS.

| Düsseldorf, *Malkesten* | London, *Athenæum* |
| Glasgow, *Western* | Rome, *Internazionale* |

Agricultural High School of Berlin
Faculty of Medicine of Paris
Faculty of Physicians and Surgeons of Glasgow
Psychological Institute of Leipzig
Royal College of Physicians, Edinburgh
Royal Institution, Edinburgh
Royal Dublin Society
Royal Society of London

DEPARTMENTS OF THE U. S. GOVERNMENT.

Bureau of Education
Bureau of Engraving
Bureau of Ethnology
Department of War
Library of Congress

National Museum
Patent Office
Smithsonian Institution
Surgeon General's Office.

INSTITUTIONS OF ART AND OF ART TRAINING.

Baltimore, Maryland Institute.
Boston, Museum of Fine Arts.
Chicago, Art Institute.
Cincinnati, Art Museum.
Milwaukee, School of Design.
Minneapolis, School of Design.
New Bedford, Swain School.
New York, Cooper Union.
New York, Metropolitan Museum of Art.
New York, National Academy of Design.
Philadelphia, Academy of Fine Arts.
Philadelphia, School of Industrial Art.
Philadelphia, School of Design for Women.
St. Louis, Museum of Fine Arts.
Washington, Corcoran Gallery of Art.

INSTITUTIONS OF SCIENCE.

Academy of Natural Sciences, Philadelphia.
American Institute, New York.
American Philosophical Society, Philadelphia.
College of Physicians, Philadelphia.
Essex Institute, Salem.
Franklin Institute, Philadelphia.
Museum of Comparative Zoölogy, Cambridge.
Museum of Natural History, New York.
Peabody Museum of Yale College.

UNIVERSITIES AND COLLEGES.

Brown	Johns Hopkins	Nebraska	Vassar
Columbia	Kansas	New York	Vermont
Cornell	Lehigh	Pennsylvania	Wellesley
Harvard	Minnesota	Princeton	Yale

LIBRARIES.

Baltimore—Peabody	Minneapolis—Public
Boston—Athenæum	New Bedford—Public
Boston—Public	New York—Mercantile
Brooklyn—L. I. Historical	New York—State
Brooklyn Library	Pennsylvania—State
Chicago—Historical	Philadelphia Library
Chicago—Public	St. Paul—Public
Cincinnati—Public	San Francisco—Public
Denver—Mercantile	Springfield (Mass.)—Public
Harlem Library	Wisconsin—State Historical
Massachusetts—State	Worcester (Mass.)—Public

It is impossible within the limits of this appendix to record the names of the many well-known *Dilettanti*, Art Connoisseurs, Manufacturers, etc., who have acquired copies of Animal Locomotion, and it is difficult, without unjust discrimination, to select a few from among the many Eminent Men whose names and works are known all over the world and who are subscribers. Among those, however, who have honored the Author by placing their names on his subscription book—all academical and university distinctions being omitted—are the following :

ARCHITECTS, PAINTERS OR SCULPTORS.

Alma-Tadema	Faed	Marks	Roth
Armitage	Fildes	du Maurier	Rümann
Becker	Falguière	Meissonier	St. Gaudens
Begas	Fremiet	von Menzel	Schilling
Bonnat	Frith	Millais, Sir J. E.	Siemering

Boughton	Garnier	Morot	Story
Bouguereau	Gérôme	Munkacsy	Thornycroft
Bridgman	Gilbert	Orchardson	Tiffany
Burnham	Gordigiani	Ouless	Vibert
Carolus—	Gow	Parsons	Vinea
Duran	Herkomer	Passini	Villefroy
Cavelier	Hunt, Holman	Poynter	Wagner
Conti, Tito	von Kaulbach	Puvis, de Ch	Watts
Dalou	Knaus	Richardson	Ward,
von Defregger	Knight	Richmond	Wells
Detaille	Kopf	Rivière-Briton	Weeks
Dubois	Leighton,SirF.	Robert-Fleury	von Werner
Eisenmenger	von Lenbach	Rodin	Whistler
Ende	von Löfftz	Roll	Zügel.

ARCHÆOLOGISTS, AUTHORS OF ART WORKS, ETC.

von Berlepsch	von Kekule	Pulszky
Bullen	Klein	Ruskin
von Duhn	Muntz	diSambuy, Conte
Ewald	Overbeck	Smith, Gen.SirR.M.
Falke	Pietsch	Treu
Furness, H. H.	Preuner	Wolff, Albert.

ANATOMISTS, ANTHROPOLOGISTS, BIOLOGISTS, ETHNOLO-
GISTS, PALÆONTOLOGISTS, PATHOLOGISTS, PHYSIOL-
OGISTS, PSYCHOLOGISTS, ZOOLOGISTS, ETC.

Acland, Sir H. W.	Haughton	Mosso
Agassiz, A.	Heidenhain	Müller, Max
Barrier	Hering	Munk
du Bois-Reymond	Humphry	Owen, Sir R.
Bowditch	Huxley	Pasteur
Bowman, Sir W.	Jensink	Pepper W.
Braune, W.	von Kölliker	Pettigrew
Brown-Sequard	von Kries	Powell
Burdon-Sanderson	Lankester	Rabl

Cleland	Leidy	Romanes
Darwin, F.	Lubbock, Sir J.	Rückert
Exner, S.	Ludwig	Schiff
Fick	Mantegazza	Schütz
Flower	Marey	Virchow, R.
Foster	Marshall	von Voit
Galton, F.	Meyer	Wear–Mitchell
Gill	Milne–Edwards	Wood
Goode, Brown	Mivart,	Wundt
Hasse	Moleschott	von Zittell.

PHYSICISTS, ETC.

Abney	Edison	Matthiessen
Blake	Glaisher	Quincke
Blazerna	von Helmholtz	Spottiswoode
Bramwell, Sir F.	Huggins	Thomson, Sir W.
Bunsen	Langley	Vogel
Ditscheiner	Mach	Weber.

MILITARY SCIENTISTS.

Field Marshal Count von Moltke
General U. S. Grant
General W. T. Sherman
General P. H. Sheridan
General R. B. Hayes.

THE SCIENCE OF ZOOPRAXOGRAPHY.

Made Popular by Suggestive Tracings from "Animal Locomotion."

A series of FIFTY ENGRAVINGS, each of which illustrates from 12 to 15 consecutive phases of some complete movement, photographed from life.

The successive phases of each action are arranged in a circle NINE INCHES IN DIAMETER; for reduced copies of some of which see appendix A.

Printed on six-ply Bristol-board and enclosed in

A STRONG CLOTH PORTFOLIO,

size 10x12 inches; price, Five Dollars in the United States; or One Guinea in Great Britain.

Sent free of postage upon receipt of price, to any country within the Universal Postal Union.

EADWEARD MUYBRIDGE,
University of Pennsylvania,
Philadelphia, U. S. A.

Or, at 10 Henrietta Street,
Covent Garden, London.

To convert the circles of figures into a
ZOOPRAXISCOPE,
cut out the disc, and, radiating from the centre thereof, about midway from the margin, cut or stamp thirteen equidistant perforations; each an inch long, and about the sixteenth of an inch wide.

Pin the centre of the disc to a handle and revolve it in the direction of the arrow, at a distance of about twenty-four inches, in front of a mirror.

By looking through the *upper* series of perforations at the reflection of the *lower* series of figures, a semblance of the original movements of life will be seen.

The figures may be appropriately colored, and the back of the cardboard disc should be painted a dark color, or covered with a piece of dark surfaced paper before cutting the perforations.

DESCRIPTIVE ZOOPRAXOGRAPHY.

An Elementary Treatise on Animal Locomotion,

BY

EADWEARD MUYBRIDGE.

Illustrated with twelve consecutive phases—occurring during a single stride—of each of the six regular progressive movements of the horse, traced from the results of an investigation made by the Author for the University of Pennsylvania.

12 mo. bound in cloth. Price in the United States, One Dollar; in Great Britain Four Shillings and Three Pence.

Sent upon receipt of price, free of postage to any country within the Universal Postal Union.

EADWEARD MUYBRIDGE,

University of Pennsylvania,

Philadelphia, U. S. A.

Or 10 Henrietta Street,
   Covent Garden, London.